BENCH TO BEDSIDE

Diagnostic Microbiology for the Clinicians

BENCH TO BEDSIDE

Diagnostic Microbiology for the Clinicians

Editor

Nancy Khardori

Professor of Medicine and Microbiology and Molecular Cell Biology
Director, Division of Infectious Disease
Department of Internal Medicine
Eastern Virginia Medical School
Norfolk, Virginia
USA

CRC Press
Taylor & Francis Group
Boca Raton London New York

CRC Press is an imprint of the
Taylor & Francis Group, an **informa** business
A SCIENCE PUBLISHERS BOOK

Cover Credit: EVMS Marketing and Communications.

CRC Press
Taylor & Francis Group
6000 Broken Sound Parkway NW, Suite 300
Boca Raton, FL 33487-2742

Printed on acid-free paper
Version Date: 20171214

International Standard Book Number-13: 978-1-4987-9969-0(Hardback)

Library of Congress Cataloging-in-Publication Data

Names: Khardori, Nancy, editor.
Title: Bench to bedside : diagnostic microbiology for the clinicians / editor, Nancy Khardori.
Description: Boca Raton, FL : CRC Press/Taylor & Francis Group, 2017. | Includes bibliographical references and index.
Identifiers: LCCN 2017040118 | ISBN 9781498799690 (hardback : alk. paper)
Subjects: | MESH: Microbiological Techniques | Diagnostic Techniques and Procedures
Classification: LCC RB38.2 | NLM QW 25 | DDC 616.07/5--dc23
LC record available at https://lccn.loc.gov/2017040118

Visit the Taylor & Francis Web site at
http://www.taylorandfrancis.com

and the CRC Press Web site at
http://www.crcpress.com

Contents

CHAPTER 1

Introduction to Clinical/ Diagnostic Microbiology

Nancy Khardori

Robert Koch in the late 1800s, working in a highly controversial area of medicine and experimental research laid the foundation for modern day diagnostic microbiology. With the establishment of Koch's postulates, not only was the germ theory of disease validated in concept but a totally unarguable way to prove it became available for all times to come. In making a microbiological diagnosis of infectious disease, we simply fulfill the first two Koch's postulates each time for obvious reasons, but the third and the fourth postulates cannot and do not need to be fulfilled for diagnosis since a causal relationship for most diseases by animal experimentation has already been established for a majority of infectious diseases. For the rest, the association has been established by molecular methods that became possible much later.

Irrespective of the type of postulates, the initial and the most critical step in the process of managing infectious diseases is the knowledge of disease causing microbes. Unfortunately, microbiology has been labeled as a "Basic Science" in most curricula and therefore relegated to the status of "having to memorize for passing examinations". Only too late do the students going into practice of medicine realize that diagnostic microbiology is an integral part of day-to-day patient care. The clinicians managing the patient and microbiologists generating information to optimize patient care are

Professor of Medicine and Microbiology and Molecular Cell Biology, Director, Division of Infectious Disease, Department of Internal Medicine, Eastern Virginia Medical School, Norfolk, Virginia.

Molecular Koch's postulates	**Koch's postulates 1884**
1. The phenotype or property under investigation should be significantly associated with pathogenic strains of a species and not with nonpathogenic strains. 2. Specific inactivation of the gene or genes associated with the suspected virulence trait should lead to a measurable decrease in pathogenicity or virulence. 3. Reversion or replacement of the mutated gene with the wild-type gene should lead to restoration of pathogenicity or virulence.	1. The same organism must be present in every case of the disease in question and its distribution in the body should be in accordance with the lesions observed. 2. The organism must be isolated from the diseased host and grown in pure culture. 3. The isolate must cause the disease, when inoculated into a healthy, susceptible animal. 4. The organism must be re-isolated from the inoculated, diseased animal.

Adapted from Rev Infect Dis 10, suppl 2, 1988 and Microbe 1,5: 223, 2006.

connected by a "flimsy thread" in the form of the "culture and sensitivity report". The individual ineptness of patient care providers to interpret these results in the context of the given patient is enhanced innumerable times by the "collective euphoria" of knowing enough. It is the aim of the ten chapters in this book to fill a gap vital not only to patient care but to the type of and dangers posed by the microbial world within and around the human race.

As antimicrobial therapy came into being in the early to mid-1900, so did the antibiotic resistant bacteria. To quote the discoverer of Penicillin, "the microbes educated to resist penicillin and a host of penicillin fast organisms is bred out… in such a case the thoughtless person playing with penicillin treatment is morally responsible for the death of the man who finally succumbs to infection with the penicillin resistant organism. I hope this evil can be averted" Alexander Fleming, 1945.

The most significant contribution to the ever escalating epidemic of antibiotic resistance is made by the inappropriate use, misuse, overuse of antibiotics for infections and "fear of infections" to the point that after a quarter of a century since the birth of antibiotics for commercial use, we have no antibiotics to treat many types of bacteria rendered resistant to multiple antibiotics. These multiply drug resistant (MDR) bacteria affect the very types of patients that are in desperate need of antibiotics and die from infections that might have been easily treatable.

All chapters in this book have focused on the "Bench to bedside" role of diagnostic microbiology that includes the *in vitro* testing for antibiotic resistance. By giving simplified information in a reader friendly format including the use of tables allows this book to become a tool for day-to-day patient care. In focusing on the microbiological diagnosis of infectious diseases rather than "empiricism" and "antibiotic choices" tabulated in a number of "pocket books", this team of authors has attempted to bring diagnostic microbiology out from the darkness of a "Basic Science" to the well-deserved glory of an "Applied Science" and connect it to everyday patient care everywhere. Even under our best behavior, we must remember the humbling quote from Hans Zinsser, a bacteriologist and a historian during the Great Depression: "Infectious disease is one of the great tragedies of living things—the struggle for existence between different forms of life...incessantly the pitiless war goes on, without quarter or armistice-a nationalism of species against species".

I gratefully acknowledge the assistance of Ms. Alfreda Johnson in getting all the materials in the final form for publication. Her expertise and quiet confidence was needed most at the conclusion, a task I consider one of the most rewarding among my professional accomplishments.

CHAPTER 2

An Overview of Microbes Pathogenic for Humans

Eric Lehrer,[1] *James Radike*[2] and *Nancy Khardori*[2,*]

Common Pathogenic Human Bacteria and their Clinical Implications

General characteristics

All bacteria are living organisms that are classified as prokaryotes. They lack membrane-enclosed nuclei and organelles, i.e., mitochondria, Golgi complexes, endoplasmic reticulum. Additionally, most bacteria contain a peptidoglycan cell wall; the characteristics of this peptidoglycan layer are the basis for the broadest of all bacterial classification schemes, the Gram stain (Table 1).

Gram-positive bacteria include members of the Genus *Staphylococcus, Streptococcus,* and *Clostridium,* just to name a few. All Gram-positive bacteria contain a thick single layer of peptidoglycan. During the Gram staining procedure, these bacteria retain the crystal violet dye while resisting subsequent decolorization and therefore do not require counterstaining with red safranin. The result is the appearance of violet colored bacterial cells on microscopy.

Gram-negative bacteria include members of the genus *Neisseria,* genus *Pseudomonas* and family *Enterobacteriaceae,* to name a few. These bacteria contain a thin peptidoglycan layer with an outer membrane. As a result,

[1] Eastern Virginia Medical School Norfolk, VA.
[2] Division of Infectious Diseases, Department of Internal Medicine, Eastern Virginia Medical School Norfolk, VA.
* Corresponding author: khardoNM@evms.edu

Table 1. Short List of Bacteria for the Clinician.

GRAM-POSITIVE				GRAM-NEGATIVE			
Aerobes		*Anaerobes*		*Aerobes*		*Anaerobes*	
Cocci	*Bacilli*	*Cocci*	*Bacilli*	*Cocci*	*Bacilli*	*Cocci*	*Bacilli*
Staphylococcus	*Corynebacterium*	*Peptococcus*	*Clostridium*	*Neisseria*	*Enterobacteriaceae*	*Veillonella*	*Bacteroides*
Streptococcus	*Bacillus*	*Peptostreptococcus*		*Moraxella*	*Moraxella*		*Fusobacterium*
Enterococcus	*Listeria*			*Hemophillus*	*Hemophillus*		*Prevotella*
	Nocardia *Lactobacillus* *Erysipelothrix*		*Actinimyces*		*Pseudomonas*		
					Brucella *Bordetella* *Francisella* *Legionella* *Chlamydia* *Bartonella*		

this thinner layer will not hold on to the crystal violet dye as well as Gram positives do. Therefore, the subsequent decolorization and counterstaining with red safranin results in the appearance of pink colored bacterial cells on microscopy.

It is important to note that while the Gram stain is incredibly useful, it does have its shortcomings, as not all bacteria possess a peptidoglycan cell wall. Members of the genus *Mycobacterium* contain mycolic acid in their cell wall and their classification requires acid-fast staining. Members of the *Mycoplasma* genus completely lack a cell wall and cannot be identified by simple staining methods.

Bacteria can also be described by their morphology. Cocci (spherical-shaped) and bacilli (rod-shaped) are common examples. Furthermore, many bacteria assume certain configurations that can further help with classification and their role in various types of infection. For example, the most common cause of community acquired pneumonia, *Streptococcus pneumoniae* frequently appears as Gram-positive lancet-shaped diplococci; the most common cause of skin abscesses, *Staphylococcus aureus* is described as having the appearance of Gram-positive cocci in clusters.

Finally, bacteria can be described with regard to their growth behaviors in the presence of or absence of oxygen. Most aerobic bacteria are considered facultative anaerobes because they are also capable of growing in anaerobic conditions. However, anaerobic bacteria are almost always incapable of growing in the presence of oxygen and are therefore known as obligate anaerobes. Bacteria can be classified even further based on phenotypic and genotypic characteristics that are beyond the scope of this overview.

Staphylococci

The genus *Staphylococcus* is a diverse group of over two dozen species. The members of these species are all Gram-positive aerobic/facultative anaerobic cocci that are non-spore forming and arrange themselves in clusters when viewed microscopically. Members of this genus are responsible for a wide array of both local and systemic diseases in humans. As a result, they are some of the most commonly encountered microbes in healthcare settings and diagnostic microbiology laboratory.

Staphylococcus aureus is the most virulent and best studied species of this genus. According to the National Healthcare Safety Network data, *S. aureus* was the most frequently reported pathogen in healthcare-associated infections in 2009 and 2010 (Sievert et al. 2013). One of the defining features of this bacterium is its ability to produce the enzyme coagulase, which is capable of converting fibrinogen to fibrin. Therefore, a coagulase test is often utilized to differentiate *S. aureus* from other members of the *Staphylococcus* genus. Additionally, coagulase acts as a virulence factor by blocking phagocytosis by immune cells (Fry 2013). Many of the other pathogenic/opportunistic members of this genus are often referred to as "coagulase negative *staphylococci*".

S. aureus commonly colonizes the anterior nares and skin of healthy adults. These colonized individuals are not only at risk of developing an infection but are also capable of transmitting the microbe to other individuals. The main mechanism of infection transmission is via contact direct or through fomites. This is especially concerning in healthcare environments where colonized healthcare workers as well as patients can transmit the microbe to others around them.

Roughly 90% of *S. aureus* infections arise in the skin and soft tissues. *S. aureus* is the most common cause of impetigo and monomicrobial skin abscesses. It is also a relatively common co-pathogen in erysipelas, cellulitis, and necrotizing fasciitis. Systemic *S. aureus* infections develop in the remaining 10%. This microbe is capable of colonizing medical devices by adherence and production of biofilms seen after placement of vascular catheters, other indwelling medical devices, and following invasive medical procedures resulting in blood stream infection. Infective endocarditis caused by *S. aureus* usually affects the tricuspid valve and those most at risk are injection drug users. Other deep tissue infections associated with *S. aureus* include osteomyelitis, septic arthritis, pneumonia, and empyema.

In addition to coagulase, *S. aureus* is capable of producing several other virulence factors. These include enterotoxin, which is a preformed toxin that is responsible for rapid-onset food poisoning; exfoliative toxin, which is responsible for staphylococcal scalded skin syndrome; and toxic shock syndrome toxin 1, which is responsible for the development of toxic shock syndrome.

Hand hygiene and decontamination of environmental surfaces are a vital component in reducing the incidence of *S. aureus* infections. This is becoming more and more important with the increasing incidence of infections with antibiotic resistant strains of *S. aureus*—Methicillin-resistant *Staphylococcus aureus* (MRSA), Vancomycin intermediate susceptible *S. aureus* (VISA), and Vancomycin-resistant *Staphylococcus aureus* (VRSA).

The coagulase-negative *staphylococci* (CNS) are common opportunistic microbial pathogens. *Staphylococcus epidermidis* is a predominant constituent of normal/resident skin flora; it is therefore a common contaminant in blood cultures. Additionally, it is capable of producing adherent biofilms, which cause infections of implantable prosthetic devices (e.g., mechanical heart valves). Another commonly seen CNS is *Staphylococcus saprophyticus*, which is a common cause of urinary tract infections in sexually active young females. Other clinically relevant CNS includes *S. lugdunensis*, which can cause skin and soft tissue infections as well as bacteremia and infective endocarditis. *S. hominis*, *S. haemolyticus*, *S. warneri*, and *S. simulants* are pathogenic in humans; however, are not commonly isolated as such. *S. pettenkoferi* is a relatively recently discovered species of this genus and has been implicated in bloodstream infections and osteomyelitis (Trulzsch et al. 2002, Loiez et al. 2007, Mammina et al. 2011).

Streptococci

The *Streptococci* are a diverse group of Gram-positive facultative anaerobic cocci that take on chain formation as viewed microscopically. This genus is also quite diverse and is responsible for a multitude of different types of infections. *Streptococci* are classified on multiple levels including their hemolytic reactions when cultured on blood agar, serologic specificity of their cell wall (Lancefield Classification), and biochemical properties.

Streptococcus pyogenes has Group A Lancefield Classification and is β-hemolytic. This bacterium is responsible for infections of the respiratory tract, e.g., pharyngitis, skin, and soft tissues; such as cellulitis. It is the most common cause of necrotizing fasciitis. *S. pyogenes* contains many virulence factors, the most unique being the M protein. A process known as *molecular mimicry* can occur in which the body generates antibodies against the M protein, which leads to the rare development of acute rheumatic fever 2–3 weeks after the resolution of a Group A *Streptococcus* (GAS) infection. Based on a similar mechanism certain strains of GAS, termed *nephritogenic* can lead to poststreptoccocal glomerulonephritis, which can present as a severe form of nephritic syndrome.

Streptococcus agalactiae, which has Group B Lancefield Classification, is β-hemolytic and another clinically significant species of the *Streptococcus* genus, especially in neonates. Group B *Streptococcus* (GBS) frequently colonizes the genitourinary and gastrointestinal tracts of humans. As a result, vertical transmission of the bacterium can occur after the onset of labor or rupture of the fetal membranes (Schrag et al. 2002). Pregnant females are screened for GBS colonization and are treated with intrapartum antibiotics to prevent transmission. If transmission from the mother to the fetus does occur, severe disease can develop such as meningitis and bacteremia. Pregnant women are often asymptomatic; however, manifestations include urinary tract infection, endometritis and invasive maternal GBS infections that can cause loss of pregnancy and preterm delivery (Regan et al. 1996, Krohn et al. 1999, Zaleznik et al. 2000, Phares et al. 2008). GBS has also been identified as a causative agent of bacteremia, sepsis, and soft tissue infections in nonpregnant adults; individuals with chronic medical conditions, such as diabetes mellitus, cancer, HIV, hepatic, and renal diseases (Farley et al.

1993). Furthermore, elderly patients and those who are residing in nursing facilities are at an even greater risk (Henning et al. 2001).

Group C and Group G *Streptococci* are commensals of the upper respiratory tract in humans and they often colonize other areas, such as the skin and gastrointestinal tract. Both of these *streptococci* are now treated as a single species—*S. dysgalactiae* subsp. *equismilis*. This bacterium shares many of the same virulence factors of *S. pyogenes* and causes similar types of infectious presentations.

The Viridans *streptococci*, are α-hemolytic and include *S. mitis*, *S. sanguis*, and *S. salivarius*. These bacteria are normal commensals of the oral cavity and are the most common cause of endocarditis in previously damaged heart valves. Additionally, *S. mutans* is the most common organism associated with dental caries. This is due to its ability to produce polysaccharides from dietary sugars that facilitate adherence of the bacteria to the teeth (Johnson et al. 1977).

Streptococcus pneumoniae is one of the most ubiquitous microbes in all of clinical microbiology. It is α-hemolytic and is a major cause of otitis media, pneumonia, bacteremia, and meningitis in both children and adults. The bacterium frequently colonizes the respiratory mucosa and its main virulence factor is an antiphagocytic polysaccharide capsule. There are close to 100 different characterized *S. pneumoniae* polysaccharide capsules, which serve as the basis for the classification of the different serotypes.

Streptococcus bovis, which includes several species that belong to Group D *Streptococci*, grow as nonhemolytic colonies on blood agar. This and related species of *streptococci* are known to cause bacteremia and infective endocarditis in adults. There is an association between infections caused by these species and gastrointestinal malignancies.

Enterococci

Enterococci (formerly known as Group D *Streptococci*) belong to the genus Enterococcus are a part of the normal flora of the human gastrointestinal tract. They are known to be a causative agent of infective endocarditis and urinary tract infections and have become the third most commonly isolated bacteria in nosocomial infections (Schaberg et al. 1991, Emori and Gaynes 1993). Two very clinically relevant species of *Enterococci* are *E. faecalis* and *E. faecium*. Antibiotic resistance is inherent among enterococci; Vancomycin resistance is particularly common in *E. faecium*. Prior to widespread emergence of resistance to Vancomycin among *Enterococci*, 85–90% of *Enterococcus* isolates were *E. faecalis*; however, the incidence of *E. faecium* has subsequently increased and now accounts for almost 40% of isolates (Hidron et al. 2008).

Anaerobic Gram-positive cocci

Peptococci are obligate anaerobic Gram-positive cocci, which are normal constituents of the flora of the mouth, upper respiratory tract, and large bowel. They are rarely the cause of soft tissue infections and bacteremia.

Peptostreptococci are obligate anaerobic Gram-positive cocci, which are part of the normal flora of the oral cavity, gastrointestinal tract, genitourinary tract, and skin. These bacteria are clinically significant, as they are a part of a group of bacteria that are the causative microbe in aspiration pneumonia.

Aerobic Gram-positive bacilli

Members of the *Bacillus* genus are notable for their ability to produce spores and their morphological appearance of chains under microscopy. *Bacillus anthracis* is the bacterium responsible for anthrax, which because of its air borne transmissibility has been used as an agent of bioterrorism. Anthrax can be transmitted by contact, via inhalation, and via the gastrointestinal tract. The cutaneous form of the disease tends to be the least severe; however, all 3 forms are capable of causing systemic disease and death. *Bacillus cereus* is known to be a cause of food poisoning due to the ingestion of preformed spores, which are able to survive in the harshest of environments. A common manifestation of *B. cereus* food poisoning is after the ingestion of reheated rice.

Corynebacterium diphtheriae, the causative organism of diphtheria is the pathognomonic member of the *Corynebacterium* genus; they do not form spores. This bacteria produces a toxin that acts as a protein synthesis inhibitor. Patients often present with a gray pseudomembrane covering the posterior oropharynx, which can spread anywhere along the respiratory tree. Additionally, patients may develop myocarditis and neurologic symptoms. The widely available toxoid vaccine has led to significant decrease in this potentially fatal infection.

Listeria monocytogenes is the most common member of the *Listeria* genus, consisting of facultative intracellular, motile, non-spore forming, and Gram-positive bacilli. This pathogen is commonly associated with foodborne infections, as it is able to replicate at refrigeration temperatures. Immunocompromised individuals (especially those with cell-mediated immune deficiencies), neonates, pregnant females, and the elderly are at an increased risk of infection with *L. monocytogenes*, manifesting as meningitis and septicemia.

Nocardia asteroides is the most commonly identified member of the *Nocardia* genus, which is responsible for nocardiosis. It is important to note that in addition to being non-spore-forming aerobic gram-positive bacilli, members of this genus stain weakly acid fast. Nocardia species are not transferred from person to person, rather they are found living in the soil and water supply. Nocardiosis is a severe infection that is commonly considered to be opportunistic; however, approximately one-third of patients with the infection are immunocompetent (Beaman and Beaman 1994). Nocardiosis is a potentially devastating infection, which is capable of infecting any organ in the human body.

Erysipelothrix rhuisopathiae is a pleomorphic, non-spore forming, gram-positive bacillus that is capable of causing local soft tissue and systemic infections. This bacterium infects many different domestic and marine animals, and human infections are often due to an occupational exposure (Wang et al. 2002).

Anaerobic Gram-positive bacilli

Members of the genus Clostridium are anaerobic bacteria that are capable of producing spores. They are found in a variety of environments, such as the gastrointestinal tract of humans and in the soil.

Clostridium botulinum is the bacterium that causes botulism. The bacteria produce a toxin that affects neurotransmission and the neuromuscular junction in musculoskeletal cholinergic nerve fibers. As a result, patients present with symmetric neurologic deficits. Most cases of adult botulism occur due to consumption of canned foods containing the preformed *C. botulinum* toxin. Interestingly, *C. botulinum* toxin has found use in some neurologic disorders and in cosmetic applications.

Clostridium tetani is the bacterium responsible for tetanus. This bacterium produces a toxin that prevents the action of inhibitory neurotransmitters at the neuromuscular junction. As a result, patients present with spasticity and rigidity. A toxoid vaccine is widely used and has been very effective in preventing outbreaks of tetanus.

Clostridium perfringens is the bacterium responsible for gas gangrene, via the production of its α-toxin, which is responsible for fermentation of cell membranes lipids. Additionally, it is capable of causing diarrhea mediated by the production of an enterotoxin.

Clostridium difficile is a commonly encountered pathogen, especially in healthcare settings. It is a normal constituent of the human GI tract; however, in patients who are receiving antibiotics, the *C. difficile* proliferates and causes pseudomembranous colitis. Diagnosis is made by detection of the *C. difficile* toxin in the stool, and handwashing with soap and water is most effective in preventing its transmission.

Members of the *Actinomyces* species are non-spore forming bacilli that can either be facultative or obligate anaerobes. They are morphologically similar to *Nocardia*; however, they do not stain acid-fast. The most clinically significant member of this genus is *A. israelii*, which is responsible for the majority of cases of actinomycosis. This bacteria is part of the normal human oral flora; however, they usually become invasive when normal oral mucosal barriers become breached. Actinomycosis presents with multiple abscesses that are connected by sinus tracts, which are often found in the cervicofacial area and are commonly associated with dental infections. Additionally, actinomycosis can manifest as central nervous system, thoracic, abdominal, and pelvic infections (Kwartler and Limaye 1989).

Aerobic Gram-negative cocci

Members of the *Neisseria* genus are aerobic or facultative anaerobic gram-negative diplococci. The two most notable members are *N. meningitidis* and *N. gonorrhoeae* that cause significant pathologic conditions in humans.

N. meningitidis is an encapsulated bacteria that is responsible for meningococcal meningitis. It is transmitted by respiratory droplets, and its ability to infect humans is dependent on its ability to adhere to respiratory epithelial cells. There are several serogroups of these bacteria, which are classified based on the capsular polysaccharide. Eight of these identified serogroups are known to cause disease in humans—A, B, C, X, Y, Z, W135, and L. There is a quadrivalent polysaccharide vaccine available, which covers the A, C, Y, and W135 serotypes, which the CDC presently recommends be administered in two doses between the ages of 11–18 in all individuals. Due to its nature as an encapsulated organism, vaccination is also recommended for individuals who have undergone a splenectomy or who have functional asplenia (e.g., sickle cell disease and hereditary spherocytosis). Individuals with deficiencies in terminal complement proteins (i.e., C5–C9) require the vaccination as well. Finally, individuals at risk for increased exposure or traveling to endemic areas (i.e., sub-Saharan Africa) should be vaccinated. Due to its highly infectious nature, droplet precautions should always be utilized for the first 24 hours when any patient suspected of having *N. meningitidis* meningitis is undergoing treatment; additionally, post-exposure antimicrobial therapy is indicated in certain cases with close contact (Siegel et al. 2007).

N. gonorrhoeae is a very common cause of sexually transmitted infections, such as urethritis, cervicitis, and pelvic inflammatory disease (PID). In these instances it commonly presents as a purulent painful discharge from the genitalia. PID is usually a later manifestation and often presents in females who have not undergone treatment or who have a long history of asymptomatic gonococcal infection. PID has been associated with increased incidence of ectopic pregnancies and other complications. The diagnosis can be confirmed via a nucleic acid amplification assay; Gram staining is also sometimes used for urethral specimens, which will characteristically show neutrophils with intracellular gram-negative cocci. *N. gonorrhoeae* can have systemic effects, such as disseminated gonococcal infection, which is associated with septic arthritis and other manifestations.

Moraxella catarrhalis is an aerobic gram-negative diplococcus that is commonly seen as a colonizer of the upper respiratory tracts of children. As a result, most children will have an upper respiratory infection with *M. catarrhalis* as the causative pathogen sometime during their childhood (Vaneechoutte et al. 1990). Adults are far less commonly colonized by this pathogen; however, *M. catarrhalis* is a commonly identified organism in acute exacerbations of chronic obstructive pulmonary disease.

Anaerobic Gram-negative cocci

This is a particularly small group of bacteria with the most notable members being of the *Veillonella* genus. *Veillonella* species rarely cause infections in humans and are normal constituents of the mouth, gastrointestinal tract, and vaginal tract (Rovery et al. 2005).

Aerobic Gram-negative bacilli

The *enterobacteriaceae* is a large family of aerobic/facultative anaerobic bacilli that do not form spores. This family consists of a diverse group of bacteria, some of which are commensals in humans and others which are purely pathogenic.

E. coli, Proteus, Enterobacter, Klebsiella, Morganella, Providencia, Citerobacter, and *Serratia* are all members of the normal intestinal flora and are often found in the upper respiratory and genital tracts. Additionally, these bacteria can be pathogenic and are often transmitted from person-to-person. Many of these bacteria cause infections in both the community and hospital setting, and are often isolated in patients with urinary tract infections, pneumonia, wound infections, and bacteremia.

E. coli has multiple strains, some of which are capable of producing different toxins, responsible for their pathologic manifestations. Enterotoxigenic *E. coli* (ETEC) is the strain that is responsible for Traveler's Diarrhea; it produces two toxins that cause a watery secretory diarrhea. Another type is Enterohemorrhagic *E. coli* (EHEC), which is commonly associated with the O157:H7 serotype. This strain produces a toxin, similar to the one seen in *Shigella,* which can lead to hemolytic uremic syndrome.

Proteus species are commonly isolated in urinary tract infections and are associated with the formation of ammonium magnesium phosphate (struvite) kidney stones and can lead to the development of large staghorn calculi in the kidneys.

Enterobacter, Serratia, and *Klebsiella* species are commonly associated with hospital-acquired infections and are often seen in immunocompromised individuals.

Shigella is a gastrointestinal pathogen that is acid-stable, meaning small doses of the bacteria are needed to cause an infection, as it is able to survive the harsh acidic environment of the human stomach. Studies have shown that as little as 10–100 organisms are enough to cause a symptomatic infection with *Shigella*, while as many as 10^5–10^8 organisms are needed to cause an infection with *Salmonella* or *Vibrio*, as these organisms are not acid stable (Bennish 1991). *S. sonnei* usually causes a mild form of watery diarrhea, while *S. dysenteriae* or *S. flexneri* cause significant symptoms of dysentery (bloody diarrhea). Proper handwashing practices are absolutely vital in preventing the transmission of *Shigella* in particular, after going to the bathroom, changing diapers, and before preparing foods. It is very important to teach and encourage effective handwashing practices among children to prevent disease transmission. Careful attention should also be paid to not permitting children to stay in daycare settings when they are experiencing diarrhea.

Salmonella is a very well-known pathogen with many species. While there are different ways to classify them, for the purpose of this discussion, it would be best to discuss them from the perspective of their associated infections. First are the species that cause typhoid (enteric) fever. Members of this genus that are responsible for typhoid fever, include *S. typhi* and *S. paratyphi*. Typhoid fever causes a severe systemic infection and is almost always spread by contact with an infected person. It has become uncommon in the United States; however, it is still a concern when traveling to endemic areas. An effective vaccine is indicated when traveling to such areas. The second group comprises the broader nontyphoidal members of this genus, which include *S. enteriditis*. These are responsible for gastrointestinal illnesses, which are commonly associated with foodborne outbreaks. Common culprits are ingestion of undercooked meat or poultry products, as well as contaminated foods. Handwashing is very important in preventing infections with *Salmonella* species; additionally, individuals with an active infection should not prepare food for others.

Pseudomonas aeruginosa is a gram-negative aerobic bacillus that is pathognomonic of healthcare-associated infections and antimicrobial resistance. It thrives in damp environments and is a common contaminant of hot tubs and contact lens solutions, largely due to its ability to produce biofilms. It is commonly isolated in burn wound infections and is often implicated in ventilator-associated pneumonia in intensive care settings. It is one of the most frequently encountered microbes in respiratory infections in patients with cystic fibrosis. *P. aeruginosa* can also cause osteomyelitis, bacteremia, malignant otitis externa ("swimmer's ear"), and diabetic wound infections. Early diagnosis and aggressive treatment are vital, especially in healthcare settings and the immunocompromised patients.

Haemophilus species are facultative anaerobic pleomorphic gram-negative cocci/cocco bacilli, which occur in pairs or short chains. *H. ducreyi* is the pathogen responsible for chancroid, a sexually-transmitted infection. *H. influenzae* compromise a group of multiple serotypes, which can be encapsulated (type a–f) or unencapsulated (nontypeable). These bacteria are spread via respiratory droplets and contain virulence factors that allow them to adhere to the respiratory mucosa. The most clinically-relevant serotype is *H. influenzae type b* (Hib), which was a common cause of bacteremia, meningitis, epiglottitis, and other serious infections prior to the introduction of the Hib polysaccharide vaccine for children. Additionally, since this bacterium is encapsulated, vaccination is vital in patients who have undergone splenectomy, have functional asplenia, or terminal complement deficiency. The other encapsulated serotypes and unencapsulated strains are frequently responsible for upper respiratory tract infections.

Brucella are small, nonmotile, facultative intracellular aerobic bacilli responsible for the zoonotic infection, brucellosis. Brucellosis is the most common zoonotic infection worldwide (Pappas et al. 2006). It is transmitted via contact with fluids from infected animals or contaminated food products (e.g., unpasteurized dairy products). Brucellosis is a systemic infection and can present with many different clinical features, such as fever, night sweats, malaise, anorexia, arthralgia, fatigue, weight loss, and depression. Infected individuals can develop a chronically infected state, which is characterized by localized infections, such as abscesses, osteomyelitis, and uveitis. Finally, pregnant women who develop brucellosis are at risk of premature birth, spontaneous abortions, and intrauterine infections with fetal death (Khan et al. 2001).

Bordatella pertussis is the pathogen responsible for pertussis. Commonly known as "whooping cough". Prior to the availability of a vaccine, this was a frequently seen illness in children. It presents with a loud cough and post-tussive emesis. Prophylaxis is indicated for close contacts of affected individuals and is usually done with a macrolide antibiotic. Unfortunately, there has been a resurgence of pertussis cases over the past several years due to waning of immunity and widespread anti-vaccination efforts. A booster dose for all adults is now recommended.

Legionella species are aerobic, gram-negative bacilli; *L. pneumophila* is the most frequently encountered species. This bacteria is responsible for causing two important clinical syndromes—Legionairre's Disease, which presents as pneumonia with high fever, and Pontiac Fever, which is an acute febrile illness that is self-limited with negligible respiratory symptoms. *Legionella* is known to thrive in damp environments and water distribution systems, such as supermarket mists and air conditioners.

Chlamydia species can be gram-negative or gram-variable and are obligate intracellular bacteria. The most notable members of this genus are *C. trachomatis, C. pneumoniae,* and *C. psittaci;* as these are the pathogens, which most commonly cause disease in humans. *C. trachomatis* is responsible for a ubiquitous sexually transmitted infection (STI). This infection is frequently asymptomatic; however, it most commonly causes urethritis in men and cervicitis in women. Due to its commonly asymptomatic presentation, untreated *C. trachomatis* infections in females can lead to pelvic inflammatory disease, consequently causing infertility and increased risk for ectopic pregnancy. Furthermore, when newborns are passing through the vaginal canal during birth in an infected mother they can develop a bacterial conjunctivitis and/or atypical pneumonia from this pathogen. *C. pneumoniae* and *C. psittaci* are commonly associated with atypical pneumonia.

Mycoplasma pneumoniae is a facultative anaerobic bacillus that lacks a cell wall, rendering it invisible on Gram staining. It is commonly transmitted via respiratory droplets between individuals in close contact, e.g., prison cellmates and military recruits. It is commonly associated with the so-called "walking pneumonia", which is considered an atypical pneumonia that often appears more severe on imaging than the actual clinical manifestations. Other common manifestations include pharyngitis, rhinorrhea, cough, and ear pain.

Rickettsia species are obligate intracellular pathogens that are transmitted by ticks. Rocky Mountain Spotted Fever (RMSF) is a serious but curable tick-borne illness. It is caused by *Rickettsia rickettsii*. RMSF has a wide clinical spectrum, from the mild to severe fulminant infections. In the early stages of the illness, most patients present with nonspecific signs and symptoms (fever, headache, malaise, myalgias, nausea), and a rash develops after several days. RMSF cases have been documented throughout North and South America. In the United States, it is most prevalent in the southeastern and south central states. *Orientia tsutsugamushi* (formerly known as *Rickettsia tsutsugamushi*) is the organism responsible for scrub typhus, the vector of which is the larval-stage trombiculid mite or chigger. Scrub typhus is endemic in many southeast Asian nations and portions of Australia that border the Indian Ocean. Scrub typhus can present either as a nonspecific febrile illness or with multisystem organ dysfunction. Additionally, given its ability to present as a hemorrhagic fever, scrub typhus should always be considered along with leptospirosis, malaria, and dengue fever in patients presenting with hemorrhagic fever.

Anaerobic Gram-negative bacilli

Bacteroides species are anaerobic gram-negative bacilli and are a major constituent of normal human bowel flora; approximately 25% of the anaerobic bacteria found in the bowel are members of the *Bacteroides* genus (Salyers 1984). *B. fragilis* is a particularly virulent member of this genus. It is the most commonly isolated pathogenic anaerobic bacteria and is the most common anaerobic component in intra-abdominal infections (Wexler 2007).

Spirochetes

Treponema pallidum is the spirochete responsible for the sexually transmitted infection, syphilis. While the overwhelmingly common mode of transmission for syphilis is through sexual contact, transfer across the placenta is the second most common (Singh and Romanowski 1999). The clinical manifestations of this infection all stem from ability of *T. pallidum* to invade blood vessels and cause a vasculitis and endarteritis. Primary syphilis is the first clinical manifestation and presents with a painless chancre at the site of inoculation that heals within a few weeks. Roughly 25% with an untreated primary infection will develop a secondary infection, usually anywhere between 2–12 weeks later, which is a systemic infection that presents with a rash, fever, malaise, anorexia, and diffuse lymphadenopathy. Individuals who are asymptomatic but have positive serologies for *T. pallidum* are said to have "latent syphilis", the communicability of infection between individuals decreases as the latent period progresses. Tertiary syphilis presents many years to decades after the primary infection and is associated with significant disease of the cardiovascular and nervous systems. Finally, syphilis can be transmitted vertically, affecting the fetus and newborns.

Borrelia burgdorferi is the bacteria responsible for Lyme disease, a tick-borne illness. There are three distinct clinical stages of Lyme disease; early localized disease is characterized by the presence of the ubiquitous targetoid skin lesion, erythema migrans (EM) and constitutional symptoms may be present. These present within 1 month following the tick bite. Early disseminated disease is characterized by multiple EM lesions and neurologic and/or cardiac abnormalities seen weeks to months after the tick bite. Finally, late Lyme disease is associated with arthritis and neurologic symptoms. Lyme disease is transmitted by the Ixodes tick, and both deer and mice serve as the main animal reservoirs. It is endemic in the northeastern and midwestern parts of the United States, as well as parts of Asia and Europe.

Members of the ***Leptospira*** genus are spirochetes that are responsible for Leptospirosis (Weil's Disease), which is found in both tropical and temperate regions; however, the incidence in the tropics is tenfold that of temperate areas (Hartskeerl et al. 2011). The natural host for these bacteria are mammals and humans are incidentally infected. The disease can manifest in many different ways, ranging from a subclinical illness to a potentially fatal systemic illness accompanied by multisystem organ failure. Humans usually become infected after exposure to environmental sources, such as animal urine, contaminated water or feces through damaged skin, mucosal barriers, or conjunctivae. Vaccination for domestic animals is strongly recommended and effective; however, a vaccine does not presently exist for humans. Humans are advised to avoid potential infectious sources and to prevent food contamination.

Mycobacteria

Mycobacteria are rod-shaped, aerobic bacteria that do not readily stain with conventional Gram-staining methods. However, when they are stained, they are capable of resisting decolorization with alcohols or acids, which is why they are termed "acid-fast" bacteria. There are many members of this genus, such as *M. tuberculosis*, the pathogen responsible for tuberculosis; *M. leprae*, the pathogen responsible for leprosy; and several other members that are responsible for various infections. While Mycobacteria most frequently infect the immunocompromised, on occasion they can cause disease in the immunocompetent.

Tuberculosis (TB) is a ubiquitous infection that is seen throughout the world and continues to be a major cause of death. It is capable of infecting any organ system in the body and classically presents with fatigue, fever, and weight loss. Pulmonary involvement is quite common and in advanced stages this manifests as chronic cough and hemoptysis. If the organism is able to disseminate into the bloodstream and spread hematogenously, military TB occurs which can be fatal. Latent TB occurs when the bacteria are present in the body but do not cause clinical symptoms; however, this form of TB can reactivate many years later. Individuals with latent TB have a roughly 10% chance to convert to an active form of infection at some later time in their lives. However, individuals with latent TB that also have HIV injections have a chance of activation up to 5–10% per year. The human host is the natural reservoir for *M. tuberculosis* and human to human transmission occurs via inhalation of respiratory droplets. Therefore, airborne precautions are instituted in healthcare settings whenever a case of TB is suspected or confirmed. The most efficacious way of preventing TB infection is via rapid diagnosis and treatment of both active and latent cases. The BCG vaccine, which is a live TB vaccine, is derived from a strain of *Mycobacterium bovis*. Research has shown the BCG vaccine to have a wide range of effectiveness and it is still used in some endemic areas. Individuals who have received the BCG vaccine will have a positive (PPD) skin test, which decreases in size with time.

Leprosy is an infection caused by *Mycobacterium leprae*, which tends to infect the skin and peripheral nerves; it is also known as *Hansen's disease*. It continues to be a major global public health concern and rapid diagnosis with adequate intense treatment is absolutely vital in order to prevent lifelong morbidity. While leprosy is endemic in tropical nations, its incidence has markedly decreased since the introduction of multidrug therapy in the early 1980s (Lastoria and Abreu 2014). It is believed that the main mode through which leprosy is transmitted is by close contact with an infected individual via respiratory droplets (Shepard 1962, Job 1990, Martins et al. 2010). Cases of transmission via skin erosions, blood, vertical transmission, breast milk, and insect bites have all been documented (Pedley 1967, Melsom et al. 1981, Job 1990, Santos et al. 2001, Ghorpade 2002, Lastoria and Abreu 2014). Humans are the main reservoirs for

Table 2. Other Pathogenic Mycobacteria.

Mycobacterium	Clinical Manifestations
M. bovis	Tuberculosis-like illness; member of the *M. tuberculosis* complex; BCG vaccine component
M. africanum	Tuberculosis-like illness; member of the *M. tuberculosis* complex
M. microti	Tuberculosis-like illness; member of the *M. tuberculosis* complex
M. avium-intracellulare	Disseminated infection usually seen in immunocompromised; member of the Mycobacterium Avium Complex
M. chimaera	Disseminated infection usually seen in immunocompromised; member of the Mycobacterium Avium Complex
M. abscessus	Pulmonary, skin, and soft tissue infections
M. chelonae	Infections of the skin and soft tissues
M. fortuitum	Skin and soft tissue infections
M. kansasii	Pulmonary infections
M. marinum	Skin nodules associated with exposure to fresh or saltwater; seen in individuals who clean fish tanks

M. leprae, but animals, such as armadillos, chimps, other species of apes, soil, water, and some arthropods have also been reported as reservoirs in the literature (Donham and Leininger 1977, Walsh et al. 1981, Kazda et al. 1986, Deps et al. 2007, Lastoria and Abreu 2014). The BCG vaccine can be used for both Leprosy and Tuberculosis with effectiveness of roughly 50% (El Lakkis and Khardori 2014).

Common Pathogenic Human Viruses and Their Clinical Implications

Viruses are very small obligate intracellular parasites that contain a genome composed of DNA or RNA surrounded by a protein coat. Once the virus is inside host cells it is able to proliferate and produce its Human Herpes Viruses infectious progeny.

There are eight human pathogens in the *herpes virus family,* including herpes simplex virus type 1 (HSV-1), herpes simplex virus type 2 (HSV-2), varicella-zoster virus (VZV), cytomegalovirus (CMV), Epstein-Barr virus (EBV), and human herpes viruses 6, 7, and 8 (HHV-6, HHV-7, HHV-8). Once an individual is infected with a herpes virus, the infection remains for the rest of their lives. These viruses are capable of having long latent periods between manifestations of clinical symptoms. Additionally, due to the inherent fragility of their viral envelopes, infection can only occur due to direct contact with the mucosal surfaces or secretions of an infected person (El Lakkis and Khardori 2014).

HSV-1 and ***HSV-2***: these viruses are found in skin and mucosal lesions, as well as in multiple types of body fluids, such as saliva and vaginal secretions. Both have the ability of infecting the oral and genital mucosa. Most commonly, HSV-1 is spread via mouth to mouth contact or by transfer of virus particles from the hands, and HSV-2 is more commonly transferred via sexual contact (El Lakkis and Khardori 2014).

In primary herpetic gingivostomatitis, clear vesicular lesions develop first and are followed by shallow ulcerations. This infection usually presents on the lips and is capable of spreading to all areas of the mouth and oropharynx. The virus then becomes latent within the trigeminal ganglion and when it reactivates it produces what are commonly known as "cold sores".

About 70% to 90% of genital herpes cases are attributed to HSV-2, while the remaining 10% are attributable to HSV-1 (El Lakkis and Khardori 2014). The primary infection does not usually produce any symptoms; however, painful lesions can develop on the shaft and glans of the penis in men and on the vulva, vagina, cervix, and perianal area in women. This virus remains dormant in the sacral ganglion and secondary episodes due to reactivation are often milder. It is important to note that patients who experience a reactivation of genital herpes often experience a prodrome of a burning sensation in the area that will develop the vesicles. Finally, it is important to note that since HSV infection is lifelong that these individuals are able to transmit the disease to their partners, regardless of the presence or absence of clinical symptoms.

VZV: these viruses produce two very commonly encountered infections. First is the primary infection, known as chickenpox (varicella) that is commonly seen in childhood. Second, is the reactivation of this infection within the sensory ganglia, known as herpes zoster (shingles) that occurs later in life.

VZV is a highly contagious virus and more than 90% of the adult population of the United States has antibodies against protein products of VZV, regardless of history of prior symptomatic infection (El Lakkis and Khardori 2014). Additionally, approximately 90% of household contacts will get varicella if they have not received the vaccine (El Lakkis and Khardori 2014). The infection can be transmitted via respiratory droplets or from contact with ruptured skin vesicles. Varicella infection in immunocompromised individuals is particularly severe with a high mortality rate. A live attenuated vaccine exists, of which two doses are recommended for children and seronegative adolescents and adults.

After primary infection or chicken pox, VZV migrates to sensory ganglia, where it remains dormant. Reactivation can occur at any time, but usually many years later, known triggers are stress and immunosuppression. Due to its residence in a sensory ganglion, shingles presents as a burning, painful vesicular rash in a dermatomal distribution. A chronic painful condition can remain after the resolution of visible symptoms, which is known as post-herpetic neuralgia. A vaccine is now used for adults to reduce the incidence of shingles.

The ***Epstein-Barr Virus*** (EBV) is responsible for infectious mononucleosis and multiple types of B-cell lymphomas. The virus exerts its pathogenicity by infecting B-lymphocytes via the CD21 surface receptor (Nemerow et al. 1987). Infectious mononucleosis is classically known as the "kissing disease", as it is spread between individuals with close contact. It presents with fever, malaise, lymphadenopathy, pharyngitis, splenomegaly, and hepatomegaly. Up to 95% of individuals are infected with EBV and remain asymptomatic throughout their lives. While it is responsible for many types of lymphomas, it is frequently associated with Burkitt Lymphoma, which is a B-cell lymphoma that is commonly seen in Africa.

Cytomegalovirus (CMV) is a ubiquitous virus that is rampant throughout the world. Among certain populations, seropositivity can approach 100% and it is more prevalent in developing countries (Krech 1973, Ho 1990). CMV can be cultured from most body secretions and is spread via close contact. Interestingly, semen has been shown to contain the highest CMV titer of any bodily fluid (El Lakkis and Khardori 2014). Most immunocompetent adults remain asymptomatic; however, a mononucleosis-like syndrome can develop. CMV causes a much more serious infection in the immunocompromised, such as patients with AIDS and those who have undergone transplants and are on chronic immunosuppressive medications. In these individuals, CMV infection can present in many ways, such as pneumonia, retinitis, encephalitis, esophagitis, and colitis. CMV is also a very common congenital viral infection since primary infection in the mother can spread via placenta to the fetus. Possible sequelae of the type of infection include microcephaly, periventricular calcifications, hepatosplenomegaly, fetal hydrops, and growth restriction.

***Human herpes virus 6 and 7*:** these viruses are the causes of *roseola*, which is also known as exanthem subitum or sixth disease. Roseola is most commonly seen in young children and is characterized by a high fever that lasts for several days that is followed by a skin rash.

***Human herpes virus 8*:** this virus is the cause of Kaposi sarcoma, which are vascular tumors known to affect HIV-positive individuals. Additionally, it has been identified as the causative agent of the lymphoproliferative disorder—Multicentric Castleman's Disease (Liu et al. 2016).

***Herpes B virus*:** this virus is endemic in macaque monkeys and is transmitted via contact with mucous membranes. The infected macaques usually experience minimal to no symptoms (Hilliard 2007). When a human is infected with this virus, often due to a bite from a macaque or secretions invading a breached mucosal barrier, the infection is severe. Humans develop a life-threatening encephalitis and CNS dysfunction, which results in death roughly 75% of the time (Hilliard 2007, El Lakkis and Khardori 2014). Survivors of the infection develop lifelong neurologic deficits.

***Rubella*:** also known as the German measles is a viral infection that is transmitted via respiratory droplets. Infection can be asymptomatic; however, when it does cause disease, it presents with a well-known set of symptoms. Usually, the illness is mild and self-limited with a prodrome consisting of headache, malaise, fever, and lymphadenopathy. This is followed by a maculopapular rash that begins on the face and spreads downward, eventually reaching the lower extremities. As time goes on, the rash begins to coalesce, forming a flushed appearance. Classically, patients will experience a postauricular lymphadenopathy. Additionally, while the majority of cases of postnatal rubella are self-limited, documented complications include arthritis, encephalitis, and thrombocytopenic purpura (Parkman 1996).

If the virus is transmitted to fetus during the first trimester of pregnancy, congenital rubella syndrome can develop. This can result in stillbirth or spontaneous abortion. If the fetus survives to term, structural anomalies may be present, such as cataracts, patent ductus arteriosus and other structural heart defects, sensorineural deafness, developmental delay, and growth retardation.

Since the introduction of MMR vaccine, rubella outbreaks have been exceedingly rare. Vaccination rate is close to 100% in the developed world; however, it is roughly 50% in developing countries (Robertson et al. 2003). Presently, there are no chemotherapeutic modalities available for rubella infection. Immunoglobulin therapy has been used to prevent infection in exposed pregnant women; however, this has not been effective at preventing transfer of the infection to fetus and is, therefore, not recommended (Parkman 1996).

Measles: also known as *Rubeola*, presents with a several-day long prodrome of high fever, malaise, and anorexia that is followed by conjunctivitis, cough, and coryza with the eventual development of a rash. The characteristic rash usually first appears in the face and neck, as discrete erythematous patches, which begins to spread to the rest of the body and becomes more confluent. Humans are the only host for this virus and infected individuals are considered contagious for several days before the appearance of the rash and several days after the resolution of the rash (El Lakkis and Khardori 2014). Due to the highly infectious nature of the virus, immunocompromised individuals can develop severe infections. Additionally, Vitamin A deficiency has been tied to increased fatalities and incidence of measles cases worldwide (Hussey and Klein 1990, Sommer and West 1996, Hatun et al. 1995). Measles virus undergoes replication in the nose, mouth, and throat and is transmitted into the air when a person coughs or talks. The virus is viable and communicable for up to 2 hours in the air and on surfaces; therefore, transmission can occur without close person-to-person contact (El Lakkis and Khardori 2014). While measles can resolve without complications, up to 30% of infected individuals experience at least one complication, which can range from mild, such as diarrhea to severe, such as subacute sclerosing panencephalitis (SSPE), which is a degenerative CNS disease that appears several years after a resolved measles infection and is invariably fatal. The MMR vaccine has had a profound impact on reducing the incidence and fatalities in the developed and the developing world.

Mumps: is a viral infection that presents with a nonspecific constellation of symptoms consisting of low-grade fever, malaise, headache, myalgias, and anorexia that is followed by the development of parotitis, roughly 48 hours later. The disease is highly contagious and is transmitted via respiratory droplets or direct contact (Gupta et al. 2005). Complications include orchitis, oophritis, meningitis, and encephalitis; which are more likely to be seen in infected adults. The MMR vaccine has been highly effective at reducing the incidence of mumps.

Influenza: is a ubiquitous infection that occurs annually with varying extremes, usually during the late fall and winter months. The varying epidemiologic patterns for these outbreaks are due to the virus's ability to undergo frequent antigenic changes. Infection presents with fever, malaise, and mylagias with involvement of the upper and/or lower respiratory tract. The majority of cases are self-limited; however, in certain at-risk populations, the infection and its complications can be fatal, as is sometimes seen in the elderly who are more prone to developing pneumonia after an influenza infection. The infection is spread via respiratory droplets, through direct contact with the droplets of an infected individual or by touching a surface contaminated with the droplets and then transferring the infection through one's eyes or mouth.

There are two main types of influenza virus, type A and type B. Type A influenza does not only infect humans but can also infect animals, such as ducks, chickens, pigs, horses, and whales. The type B virus is only capable of infecting humans. The CDC presently recommends that all individuals > 6 months of age be vaccinated against influenza on an annual basis. Hospitalized patients with influenza should be placed on droplet precautions.

Respiratory syncytial virus (RSV): a very common virus that is responsible for respiratory infections throughout the world, especially during the winter months and in younger individuals. It is believed that by the age of 2 years most children will have experienced at least one RSV infection (Borchers et al. 2013). The most common manifestation of the virus in infants and young children is bronchiolitis. It is also a significant, but often unrecognized cause of respiratory tract infections in adults, and there is an increased risk in individuals who are immunocompromised or have pre-existing cardiopulmonary disease. The virus is usually transmitted by direct contact; however, aerosol droplets have been implicated as well (Pfaller and Herwaldt 1988). Presently, immunoprophylaxis with the monoclonal antibody Palivizumab is available for high-risk/immunocompromised infants; however, a vaccine is not yet available.

Parvovirus B19: is a very common virus that infects individuals throughout the world. It is usually acquired during childhood with decreasing incidence with advancing age. Anywhere between 70–85% of adults are seropositive from a prior parvovirus B19 infection (Cohen and Buckley 1988, Kelly et al. 2000). Individuals can either be asymptomatic or can develop non-specific symptoms, such as fever, myalgias, and fatigue. Others can present with the classic erythematous rash on both cheeks, which is known as "erythema infectiousum"; or "fifth disease". In individuals with sickle cell disease, or any type of chronic hemolytic disorders, aplastic anemia can occur in the setting of this infection. Vertical transmission during pregnancy is possible, which can result in fetal complications, such as hydrops fetal is and miscarriage. Studies have shown that proper handwashing is the most important method to prevent transmission of infection and that close contact with individuals who are experiencing respiratory symptoms or fever should be avoided as best as possible (Katragadda et al. 2013, Liefeldt et al. 2002).

Rhinovirus: these viruses were first discovered in the 1950s, they are responsible for more than 50% of all common colds and cost billions of dollars a year in clinic and time off from work (Jacobs et al. 2013). These viruses have also been linked to exacerbations of asthma and COPD. Infections usually last for 5–7 days and are self-limited. Transmission is through aerosolized infectious particles.

Adenovirus: these are very common viruses that cause several different types of self-limited infections; however, instances of fatal infections in the young and immunocompromised have been reported (Jernigan et al. 1993). They are responsible for a wide variety of infections, most commonly diseases of the upper respiratory tract. However, more serious infections, such as pneumonia, ophthalmologic, gastrointestinal, and neurologic infections are well recognized. There is no vaccine available for the general public; however, the Department of Defense recommends administration of a live oral vaccine for U.S. military personnel between the ages of 17 and 50 (El Lakkis and Khardori 2014).

Enteroviruses: this is a large family of viruses, which includes polioviruses, groups A and B coxsackieviruses, echoviruses, and enteroviruses. Infection with many of these pathogens often does not cause any clinical symptoms and when they do, it is often a self-limited febrile illness with no long-term complications. Patients can experience skin rashes, rashes of the mucous membranes, meningitis, and a myopericarditis from these viruses as well. These viruses are spread by fecal-oral contact and are seen throughout the world. Handwashing and effective sewage handling are vital for disease control. Poliomyelitis, once a severe disease that caused permanent paralysis in sufferers, has been eradicated in the United States and in many other nations worldwide since the discovery of the polio vaccine by Dr. Jonas Salk in 1955.

Human papillomavirus (HPV): there are roughly 150 known viruses that belong to this group (Bernard et al. 2010). HPV serotypes 1 and 4 are responsible for skin warts, commonly known as verruca vulgaris, which are transmitted by close physical contact. HPV serotypes 6 and 11 are responsible for genital warts, known as condyloma accuminatum. HPV serotypes 16 and 18 are associated with squamous cell carcinoma of the cervix and anus. The risk of developing squamous cell carcinoma of the cervix is roughly 250–450 times higher if someone is infected with HPV 16 or 18 versus uninfected individuals (Munoz et al. 2003, Basu et al. 2013). It is customary to divide HPV serotypes into high-risk and low-risk; the risk is determined by the microbiology of the specific serotype and its conferred ability to cause cancer. There has been an increased focus on primary prevention of cervical cancer via vaccination, which has been highly effective. Presently, there are two vaccines available, the quadrivalent vaccine (Gardasil), which covers HPV types 6, 11, 16, 18; and the divalent vaccine (Cervarix), which covers HPV types 16 and 18. Recently 5 more serotypes have been added to the quadrivalent vaccine. The CDC presently recommends vaccination for all children and adolescents, which can begin at the age of 9 and requires only 2 doses up to a year apart. They also recommend catch-up vaccinations for males and females through the age of 26.

Human immunodeficiency virus (HIV): is a retrovirus that arose from immunodeficiency viruses seen in African primates. In particular, HIV-1 was transmitted from apes and HIV-2 from sooty mangabey monkeys (Sharp and Hahn 2011, Maartens et al. 2014). HIV-1 tends to have a high degree of mutation and genetic diversity due to its error-prone reverse transcriptase enzyme. HIV-2 is mainly seen in West Africa and causes a disease similar to that caused by HIV-1. However, it usually has a slower progression and is a less transmissible virus than HIV-1 (Sharp and Hahn 2011, Maartens et al. 2014). HIV infects humans when it comes in contact with tissues that line the vagina, anus, mouth, and eyes; additionally, infection can occur when the virus enters through breached skin. The major factor that determines the risk of transmission of the virus is the plasma HIV viral load of the infected individual. Untreated HIV infection progresses through three stages. The first stage, which is known as the primary infection or acute retroviral syndrome occurs within weeks of first acquisition of the virus. It is characterized by a flu-like illness that is seen in roughly 50% of patients and resolves on its own within weeks. During this stage, the virus is replicating rapidly and individuals are highly infectious with high plasma HIV viral loads and decreasing $CD4^+$ helper T-cell counts. Once this stage resolves, those infected enter a stage of clinical latency, which is an asymptomatic stage that lasts an average of 8–10 years. The final stage is Acquired Immune Deficiency Syndrome (AIDS), which is when CD4 counts fall below 200 and patients are at a high risk for opportunistic infections, cancers, weight loss, and dementia. The most common mode of HIV transmission is through sexual contact at the genital or anal mucosa; other routes of injection include exposure to infected blood or blood products and vertical transmission from the mother to the infant. A vaccine is not available and most experts recommend post-exposure prophylaxis with three antiretroviral drugs for a duration of 4 weeks. A combination drug is now available for pre-exposure prophylaxis in high risk groups.

Hepatitis A virus (HAV): is a picornavirus, with an RNA genome that is particularly prevalent in areas with poor sanitation and is transmitted via the fecal-oral route, making adequate sanitation systems and supply of clean water vital in preventing its spread. In the most endemic areas, such as Latin America, Asia, Africa, and the Middle East, where seroprevalence rates approach 100% (Franco et al. 2012). Hepatitis A is the most common form of acute viral hepatitis worldwide and usually presents with a nonspecific prodrome that can consist of malaise, fever, weakness, and nausea that tends to improve with the onset of jaundice. Individuals are most infectious in the two weeks preceding the onset of jaundice or elevation of liver enzymes, as this is when the concentration of viral particles in greatest in the stool. Once jaundice begins to appear, the viral concentration in the stool begins to decline and patients are usually no longer infectious after 1–2 weeks (Koff 1998, Wasley et al. 2006). The vast majority of individuals experience resolution of infection without significant long-term complications. Roughly, 0.2% of patients experience acute liver failure and death with an increased risk in those who are of advanced age and have a history of chronic liver disease (Keeffe 2006). There is a vaccine available, recommendations for its use vary widely depending on one's country and travel history.

Hepatitis B virus (HBV): is a widely studied virus that can cause both acute and chronic infection. During the acute phase, varying presentations are possible, such as subclinical or anicteric hepatitis (70%), icteric hepatitis (30%), and fulminant hepatitis (0.1–0.5%) (El Lakkis and Khardori 2014). Chronic infection can range from an asymptomatic carrier state to a chronic symptomatic state with associated cirrhosis and hepatocellular carcinoma. Acute symptomatic HBV infection in adults presents with anorexia, nausea, jaundice, and discomfort in the right upper quadrant (RUQ) that can last for 1–3 months. Interestingly, roughly 5% of these acutely infected adults will develop a chronic infection; however, roughly 90% of individuals who are infected at birth will become chronically infected (Liaw et al. 1998). Studies have shown that the transmission modality of HBV varies by geographic region. Perinatal transmission is most common in areas with high prevalence and decreased access to vaccines, such as in southeast Asia and China; while transmission due to sexual contact and intravenous drug use is more common in the United States, Canada, and Western Europe (El Lakkis and Khardori 2014). Vaccination is universally recommended, as it has been shown to be highly effective. Current guidelines recommend commencing HBV vaccination at birth and completing the 3 dose series between 6 and 18 months. After HBV exposure, regardless of the context, expeditious and appropriate prophylactic intervention can prevent the onset of HBV infection and possible long-term complications. Individuals who have completed the vaccination series and have not had their titers checked should receive a single booster dose and unvaccinated individuals should receive both the vaccine and Hepatitis B Immune Globulin (HBIG) as soon as possible after exposure. The same algorithm should be used for children born to HBV positive mothers.

Hepatitis C virus (HCV): is another widely studied type of hepatitis virus. When individuals become acutely infected with HCV, they often are asymptomatic or experience negligible to mild clinical symptoms. Jaundice is present in roughly 25% of cases (El Lakkis and Khardori 2014). Unlike HBV, up to 80% of individuals who experience an acute HCV infection will become chronic carriers; additionally, the most common mode of transmission is via the percutaneous route. While sexual transmission is possible, it is uncommon; therefore, HCV-positive patients with a single long-term steady sexual partner do not need to change their sexual practices (El Lakkis and Khardori 2014). However, individuals with HIV infection (homosexual and heterosexual) or with multiple intimate partners should use condoms. Women should be counseled that roughly 6 in every 100 infants born to HCV positive mothers develop HCV infection (El Lakkis and Khardori 2014); however, breastfeeding has not been shown to directly transmit HCV infection. Presently, there is no post-exposure prophylaxis available. However, recent pharmacologic advances, starting with the 2014 introduction of Ledipasvir/Sofosbuvir have led to effective and convenient options for treatment of all HCV genotypes.

Hepatitis D virus (HDV): is a less common hepatitis virus, which is often considered to be "defective." Since it requires the presence of an HBV infection in order to replicate. Transmission routes are similar to that of HBV: parenterally, sexually, and vertically. There are two major categories of HDV infection: coinfection (concurrent transmission of HBV and HDV) and superinfection (transmission of HDV on top of a preexisting HBV infection). Generally speaking, patients have a worse prognosis and are more likely to develop a chronic carrier state when infected with both HDV and HBV, rather than with HBV alone.

Hepatitis E virus (HEV): is a small virus with an RNA genome. While it is seen throughout the world, its highest prevalence is in Africa and Asian countries. Transmission of the virus can occur through multiple modalities, such as contaminated water supplies and food, blood products, and from the mother to child. Of note, several different genotypes of the HEV exist and differ in their route of transmission. Most patients with an acute infection are asymptomatic. While acute hepatic failure is relatively rare in general population, pregnant women experience a high mortality rate when infected with HEV.

Rabies virus**:** Rabies virus infection if left untreated is almost always fatal; however, the disease can be prevented with administration of post-exposure prophylaxis and effective wound care. Initially, rabies may present as an influenza-like illness with nonspecific symptoms, which can persist for several days before the patient develops the classic symptoms associated with rabies, such as CNS dysfunction, hydrophobia, delirium, agitation, and insomnia. Transmission occurs via infected saliva or brain/nervous system tissue. Transmission of rabies usually occurs when an infected host bites an uninfected animal. Animal bites are almost always the method by which rabies is transmitted. While all mammals can become infected with the rabies virus, only certain ones are important reservoirs for the disease. In the United States, the main reservoir for rabies are bats; however, the virus has also been identified in raccoons, skunks, coyotes, and foxes. In the developing world, dogs continue to be the main reservoir for the virus. It is absolutely vital to seek medical attention immediately for post-exposure prophylaxis if any concern about being bitten by a rabid animal exists. Post-exposure prophylaxis consists of both passive (administration of the rabies immunoglobulin) and active immunization, i.e., administration of multiple doses of the rabies vaccine. Pre-exposure prophylaxis should be targeted for at-risk individuals, such as veterinarians and those traveling to endemic areas.

Flaviviruses**:** this is an important family of RNA viruses, which are all transmitted by mosquitoes, making the use of mosquito repellants, proper clothing, and mosquito nets vital to prevent their spread.

The ***Yellow Fever virus*** is the cause of yellow fever, a hemorrhagic fever that still affects many individuals in Africa and Asia. It presents with nonspecific symptoms, such as fever, myalgias, headache, and fatigue. These nonspecific symptoms generally resolve in roughly 48 hours. Infected individuals then begin to experience hepatic dysfunction, renal failure, coagulopathy, and shock. A live-attenuated vaccine exists and the CDC recommends its administration for individuals traveling to endemic areas.

The ***West Nile virus*** is the causative agent of West Nile Fever. Most infected patients remain asymptomatic. West Nile Fever is a self-limited illness that presents with nonspecific symptoms, such as fever, malaise, back pain, anorexia, and a maculopapular rash. However, patients may also experience neurologic symptoms, such as encephalitis and meningitis, which carries a higher risk of death. Wild birds serve as hosts of the virus but they usually do not display any symptoms. The virus is found throughout the world, including North America.

The ***St. Louis Encephalitis virus*** is seen throughout the Americas. While infection with the virus rarely results in clinical manifestations, serious complications, such as encephalitis are possible. This is most frequently seen in elderly infected individuals.

The ***Japanese Encephalitis virus*** infection commonly presents with an acute encephalitis; however, it can also present with aseptic meningitis or a nonspecific febrile illness and headache. Humans are incidental hosts and are not able to transmit the virus to arthropod vectors.

Table 3. Short List of DNA and RNA Viruses for the Clinician.

<table>
<tr><th>DNA Viruses</th><th>RNA Viruses</th></tr>
<tr><td>• Herpesviruses
o HSV-1 and HSV-2
o VZV
o EBV
o HHV-6 and HHV-7
o HHV-8
o Herpes B Virus
• Parvovirus B19
• Adenovirus
• HPV
• HBV</td><td>• Rubella
• Measles
• Mumps
• Influenza
• RSV
• Rhinovirus
• Enteroviruses
o Poliovirus
o Coxsackieviruses A and B
o Echovirus
• HIV
• HAV
• HDV
• HEV
• Rabies
• Flaviviruses
o HCV
o Yellow Fever
o West Nile
o St. Louis Encephalitis
o Japanese Encephalitis</td></tr>
</table>

Dengue: is a virus that is endemic throughout the world, except in Europe and Antarctica. The disease has a wide range of clinical manifestations, ranging from a self-limited milder form to a more severe form that is associated with hemorrhagic fever and shock, the latter being more common.

Prions and Their Clinical Implications

Prions are small infectious pathogens that cause rapidly fatal degenerative neurologic diseases. These pathogens are particularly noteworthy for their resistance to many standard procedures of decontamination and sterilization. Due to their lack of nucleic acid, sterilization techniques that normally cause nucleic acid destruction are not effective. Interestingly, prion diseases can undergo incubation periods up to several decades before exhibiting clinical symptoms. Once these clinical symptoms manifest, death usually follows within several months. There are five known prion diseases described in humans: Kuru, Creutzfeldt-Jakob Disease (CJD), variant Crutzfeldt-Jakob Disease (vCJD), Gerstmann-Strausslër-Scheinker Syndrome (GSS), and fatal familial insomnia. Bovine spongiform encephalopathy in humans, colloquially known as "mad-cow disease" is a prion disease that affects animals and its appearance has increased public attention to prion diseases as a whole. Prions exert their effects by causing host-encoded prion proteins to undergo conformational changes in their structure, forming aberrant proteins. These proteins accumulate over the course of decades and manifest as neurologic dysfunction with the subsequent development of dementia. Death usually ensues over the course of months. Kuru was endemic in remote tribal areas of New Zealand, believed to be due to cannibalistic practices. The disease is no longer endemic since the practice was stopped. Prion infections are transmitted to humans through ingestion of bovine meat products or through medical interventions involving infected tissues (e.g., corneal transplants).

Common Pathogenic Human Fungi and Their Clinical Implications

Candida: is a yeast that can cause local and systemic infections. While over 20 species of Candida exist, the most commonly encountered species in the healthcare setting is *C. albicans*. Although its presence is not always pathologic, such as on skin surfaces, it can cause local superficial infections, mucosal infections, and systemic infections if it enters the bloodstream. Individuals who are chronically immunosuppressed, such as those taking corticosteroids for chronic medical conditions and patients with neutropenia are at high risk of developing serious and disseminated infections.

***Aspergillus*:** is a mold, which is found throughout our natural habitat; therefore, we are exposed to it on a regular basis and the vast majority of individuals remain unaffected. Manifestations can range from mild allergic reactions to systemic disease and death. Individuals that are in immunocompromised states are at a significantly higher risk of developing systemic disease.

***Cryptococcus*:** is a yeast found in soil throughout the world and is associated with pigeon droppings; however, pigeon droppings have not been documented to be an actual source of disease transmission to humans (El Lakkis and Khardori 2014). *Cryptococcus* is known to cause fungal meningitis especially in patients with HIV infection.

The fungi that cause endemic mycoses: ***Blastomycosis*, *Histoplasmosis*, *Coccidioidomycosis***, and ***Paracoccidioidomycosis*** are all dimorphic fungi that are each uniquely endemic in different parts of the world. Infection usually occurs via inhalation of fungal particles or spores, and clinical manifestations occur in a minority of patients. If infected, regardless of symptomatology, they are not transmissible between humans.

Sporothrix schenckii is the dimorphic fungus responsible for *Sporotrichosis*. This infection classically presents ous contact with infected thorns, such as in gardeners. The result is a cutaneous infection that responds well to antifungal medications. Systemic infection is possible, but a rarer event, individuals with chronic immunosuppression and alcoholism are at an increased risk.

Members of the ***Malassezia*** genus are responsible for the infection *tinea versicolor*. This is a very common superficial fungal infection that responds well to medical treatment. It presents with hypopigmented, hyperpigmented, or erythematous macules on the trunk and proximal upper extremities. These infections are commonly found in individuals exposed to warm weather that results in perspiration, as *Malassezia* thrives in warm humid environments. *M. furfur* is the species that causes invasive infection in patients receiving intravenous high lipid formulations.

Dermatophytes, include ***Epidermophyton*, *Trichophyton***, and ***Microsporum***. These cause superficial infections of keratinized tissues, such as the skin, nails, and hair follicles. Dermatophyte infections are named by the site at which they exert their effects. Such as, tinea cruris, tinea capitis, tinea pedia, and tinea corporis. These infections can be transmitted by contact with infected soil, humans, or animals.

Common Pathogenic Human Parasites and Their Clinical Implications

Protozoa

Plasmodium: this is the genus of protozoa that are responsible for malaria. Presently, there are 5 identified species that cause the disease in humans—*P. falciparum*, *P. vivax*, *P. ovale*, *P. malariae*, and *P. knowlesi*. Malaria can be a very serious disease responsible for the deaths of roughly 2,000 people per day, mainly children in Africa (White et al. 2014). Malaria is transmitted via the bites of the female Anopheles mosquito, which usually strike between dusk and dawn. Initial symptoms tend to be extremely diverse and nonspecific, such as tachycardia, tachypnea, chills, malaise, anorexia, vomiting, and diarrhea. Subsequently, cyclical fever patterns develop, which occur from every 48–72 hours. Due to their ability to infect erythrocytes, complications result from anemia and hypoxic injury due to endothelial damage. While each of the aforementioned species is known to cause the disease, the form due to *P. falciparum* is the most aggressive. In endemic areas, young children and pregnant women are at a particularly high risk for the disease. As children age, repeated infections allow for the development of partial immunity, which results in less severe symptoms with subsequent infections. However, as individuals become older, immunity wanes and they enter a high risk state of developing severe symptoms. Prevention is very important for international travelers, and includes insect repellents and pre-exposure chemoprophylaxis.

Giardia lamblia is ubiquitous throughout the world for causing persistent diarrhea with varying degrees of severity. Stools tend to be watery and foul smelling due to the presence of steatorrhea. Infection occurs when individuals ingest giardia cysts. This is commonly transmitted via contaminated water or foods and is spread between individuals via the fecal-oral route. A common scenario is a camper who begins to experience severe diarrhea after drinking well water. Therefore, if it is important to avoid possible sources of contamination and to properly treat drinking water from wells, lakes, and streams with boiling, filtering, or iodine treatment.

Entamoeba histolytica is present throughout the world and tends to be more prevalent in areas with warmer weather and poor sanitation. Roughly 90% of infections are asymptomatic with the remainder producing a wide array of clinical symptoms (El Lakkis and Khardori 2014). Intestinal manifestations include diarrhea, flatulence, and abdominal pain; while extraintestinal manifestations result in the development of amebic abscesses most commonly seen in the liver. The most common sources of infection are contaminated drinking water and food. Transmission from person to person is via the fecal-oral route and adequate hygiene and sanitation practices are vital to prevent transmission in communities.

Other Protozoa

Protozoa/Disease	Transmission	Epidemiology	Symptoms	Prevention
Leishmania donovani/ Visceral Leishmaniasis or kala-azar	Sand Fly	Mediterranean basin and parts of China/Russia Sub-Saharan Africa India and neighboring countries + Kenya	Begins with a malaria-like illness. As organisms proliferate, they invade the liver/spleen and bone marrow, resulting in hepatosplenomegaly, anemia, and weight loss	Prompt treatment and control of reservoir hosts; protection from sand flies
Leishmania tropica complex, *L. aethiopica*/old world cutaneous Leishmaniasis	Sand Fly	Asia, Africa, and Europe	A red papule at the site of the sand fly bite that enlarges and ulcerates	Prompt treatment and elimination of ulcers; protection from sand flies
Leishmania mexican, *L. braziliensis*/new world cutaneous and mucocutaneous Leishmaniasis	Sand Fly	South and Central America – highly prevalent in the Amazon basin	Sores and lesions that can disseminate cutaneously	Avoidance of endemic areas; prompt treatment of infected individuals, protection from sand flies
Trypanosoma brucei complex/ African trypanosomiasis ("sleeping sickness")	Tsetse fly	*T. brucei gambisense* – tropical west and central Africa *T. brucei rhodesiense* – east Africa (predilection for cattle-ranching nations)	Superficial ulceration at site of insect bite, progression to lethargy, tremors, and mental disability; eventually resulting in convulsions, hemiplegia, and death	Careful control of tsetse fly breeding and use of insecticides; insect repellants and proper clothing; prompt treatment
Trypanosoma cruzi/ American trypanosomiasis (Chagas' Disease)	Triatomine insects or Kissing bugs	South and Central America	Often asymptomatic, possible erythema at site of insect bite that is accompanied by fever, myalgias, and fatigue; long-term complications are achalasia, myocarditis, and hepatosplenomegaly	Insect control; prompt treatment of infected individuals

Toxoplasma gondii/ Toxoplasmosis	Domesticated cats/felines are definitive host; humans/mammals are intermediate hosts; usually acquired by ingestion; transplacental transmission is also possible	Organisms spread to brain, lungs, liver and eyes; immunocompetent individuals are often asymptomatic; AIDS patients often develop an encephalitis; congenital infection can cause stillbirth, abortion, and neonatal disease (encephalitis, chorioretinitis, and hepatosplenomegaly)	Cooking food to safe temperatures and avoiding untreated contaminated drinking water; pregnant/ immunocompromised should avoid changing cat litter boxes or use disposable gloves and thoroughly wash hands after they are finished
Cryptosporidium parvum/ Cryptosporidiosis	Fecal-oral transmission of oocysts from human or animal sources (usually from contaminated drinking water)	Diarrhea, which is often fatal in patients with AIDS	Good hygienic practices
Trichomonas vaginalis/ Trichomoniasis	Distributed throughout the world and most commonly transmitted sexually, but can be passed via fomites as well; rarely infants can become infected when passing through the birth canal	Women experience vaginitis with vaginal discharge and dysuria; men are usually asymptomatic	Good personal hygiene and safe sexual practices

Helminths

Flukes

Helminth/Disease	Location in Host	Epidemiology	Transmission
Schistosoma haematobium/ Schistosomiasis	Venous circulation of the human bladder	Africa and Middle East	Larva penetrate the skin from snail-contaminated water
Schistosoma mansoni/ Schistosomiasis	Venous circulation of the small intestine	Africa, South America, and Middle East	Larva penetrate the skin from snail-contaminated water
Paragonimus westermani/ Paragonimiasis	Lungs, brain, and other sites	Asia and various African countries	Raw crabs and other freshwater crustaceans

Roundworms

Helminth/Disease	Location in Host	Epidemiology	Transmission
Ascaris lumbricoides/Ascariasis	Small intestine; larvae via the lungs	More than 700 million infected people worldwide	Contaminated food and water
Ancylostoma duodenale/ Hookworm infection	Small intestine; larvae via the lungs	Temperate climates	Infected soil via the skin
Necator americanus/Hookworm infection	Small intestine; larvae via the lungs	Tropics and North America	Infected soil via the skin
Ancylostoma braziliense/ Cutaneous larva migrans	Subcutaneous migratory larvae	Worldwide	Contact with dog or cat contaminated soils
Toxocara canis and *Toxocara cati*/ Visceral larva migrans	Cerebral, myocardial, and pulmonary migratory larva	Worldwide	Ingesting soil contaminated by dogs or cats
Strongyloides stercoralis/ Strongyloidiasis	Duodenum, jejunum, and larvae via the skin and lungs	Worldwide	Infected soil via the skin

Enterobius vermicularis/Pinworm	Cecum, colon	Worldwide	Direct infection from a patient (fecal-oral route)/ self-contamination
Wucheria bancrofti/Filariasis	Lymphatic tissue and vessels; microfilariae throughout the bloodstream	More than 100 million infected people worldwide	Mosquito
Onchocerca volvulus/ Onchocerciasis	Skin, lymphatic vessels, cornea; microfilariae throughout the subcutaneous tissue fluids	Africa and South America	Black fly
Loa loa/Loiasis	Conjunctiva; microfilariae throughout the bloodstream	Western and Central Africa	Mango fly and deerfly
Dracunculus medinensis/ Dracunculiasis	Subcutaneously – usually via the lower extremities	India, Nile river valley; central, wester, and equatorial Africa	Consumption of contaminated drinking water
Trichinella spiralis/Trichinellosis	Larvae present in striated muscle	Worldwide with over 100 different infected animal species	Consumption of undercooked pork products

Tapeworms

Helminth/Disease	**Location in Host**	**Epidemiology**	**Transmission**
Echinococcus granulosus (Dog Tapeworm)/Hydatidosis	Larvae present in the liver, lung, brain, peritoneum, long bone, and kidney	Worldwide	Ingestion of food contaminated by feces of dogs; handling of infected dogs
Taenia saginata (Beef Tapeworm)/ Taeniasis	Small Intestine	Worldwide	Undercooked beef
Taenia solium (Pork tapeworm)/ Taeniasis	Small Intestine	Worldwide	Undercooked pork
Diphyllobothrium latum (Broad or fish tapeworm)/ Diphyllobothriasis	Small Intestine	Worldwide	Undercooked freshwater fish

References

Basu, P., D. Banerjee, P. Singh, C. Bhattacharya and J. Biswas. 2013. Efficacy and safety of human papillomavirus vaccine for primary prevention of cervical cancer: A review of evidence from phase III trials and national programs. South Asian J. Cancer 2(4): 187–192.

Beaman, B.L. and L. Beaman. 1994. Nocardia species: host-parasite relationships. Clin. Microbiol. Rev. 7(2): 213–264.

Bennish, M.L. 1991. Potentially lethal complications of shigellosis. Rev. Infect. Dis. 13 Suppl 4: S319–324.

Bernard, H.U., R.D. Burk, Z. Chen, K. van Doorslaer, H. zur Hausen and E.M. de Villiers. 2010. Classification of papillomaviruses (PVs) based on 189 PV types and proposal of taxonomic amendments. Virology 401(1): 70–79.

Borchers, A.T., C. Chang, M.E. Gershwin and L.J. Gershwin. 2013. Respiratory syncytial virus—a comprehensive review. Clin. Rev. Allergy Immunol. 45(3): 331–379.

Cohen, B.J. and Marie M. Buckley. 1988. The prevalence of antibody to human parvovirus B 19 in England and Wales. Journal of Medical Microbiology 25(2): 151–153.

Deps, P.D., J.M. Antunes and J. Tomimori-Yamashita. 2007. Detection of Mycobacterium leprae infection in wild nine-banded armadillos (Dasypus novemcinctus) using the rapid ML Flow test. Rev. Soc. Bras. Med. Trop. 40(1): 86–87.

Donham, K.J. and J.R. Leininger. 1977. Spontaneous leprosy-like disease in a chimpanzee. J. Infect. Dis. 136(1): 132–136.

El Lakkis, Iass and Nancy Khardori. 2014. The mighty world of microbes: An overview. *In*: Wattal, C. and N. Khardori (eds.). Hospital Infection Prevention: Principles & Practices. Springer India, New Delhi.

Emori, T.G. and R.P. Gaynes. 1993. An overview of nosocomial infections, including the role of the microbiology laboratory. Clin. Microbiol. Rev. 6(4): 428–442.

Farley, M.M., R.C. Harvey, T. Stull, J.D. Smith, A. Schuchat, J.D. Wenger et al. 1993. A population-based assessment of invasive disease due to group B Streptococcus in nonpregnant adults. N Engl. J. Med. 328(25): 1807–1811.

Franco, E., C. Meleleo, L. Serino, D. Sorbara and L. Zaratti. 2012. Hepatitis A: Epidemiology and prevention in developing countries. World J. Hepatol. 4(3): 68–73.

Fry, D.E. 2013. The continued challenge of *Staphylococcus aureus* in the surgical patient. Am. Surg. 79(1): 1–10.

Ghorpade, A. 2002. Inoculation (tattoo) leprosy: a report of 31 cases. J. Eur. Acad. Dermatol. Venereol. 16(5): 494–499.

Gupta, R.K., J. Best and E. MacMahon. 2005. Mumps and the UK epidemic 2005. BMJ 330(7500): 1132–1135.

Gurung, S., J. Pradhan and P.Y. Bhutia. 2013. Outbreak of scrub typhus in the North East Himalayan region-Sikkim: an emerging threat. Indian J. Med. Microbiol. 31(1): 72–74.

Hartskeerl, R.A., M. Collares-Pereira and W.A. Ellis. 2011. Emergence, control and re-emerging leptospirosis: dynamics of infection in the changing world. Clin. Microbiol. Infect. 17(4): 494–501.

Hatun, S., T. Tezic, B. Kunak and A.B. Cengiz. 1995. Vitamin A levels of children with measles in Ankara, Turkey. Turk. J. Pediatr. 37(3): 193–200.

Henning, K.J., E.L. Hall, D.M. Dwyer, L. Billmann, A. Schuchat, J.A. Johnson et al. 2001. Invasive group B streptococcal disease in Maryland nursing home residents. J. Infect. Dis. 183(7): 1138–1142.

Hidron, A.I., J.R. Edwards, J. Patel, T.C. Horan, D.M. Sievert, D.A. Pollock et al. 2008. NHSN annual update: antimicrobial-resistant pathogens associated with healthcare-associated infections: annual summary of data reported to the National Healthcare Safety Network at the Centers for Disease Control and Prevention, 2006–2007. Infect. Control Hosp. Epidemiol. 29(11): 996–1011.

Hilliard, J. 2007. Monkey B virus. *In*: Arvin, A., G. Campadelli-Fiume, E. Mocarski et al. (eds.). Human Herpesviruses: Biology, Therapy, and Immunoprophylaxis. Cambridge.

Ho, M. 1990. Epidemiology of cytomegalovirus infections. Rev. Infect. Dis. 12 Suppl 7: S701–710.

Hussey, G.D. and M. Klein. 1990. A randomized, controlled trial of vitamin A in children with severe measles. N Engl. J. Med. 323(3): 160–164.

Jacobs, S.E., D.M. Lamson, K. St. George and T.J. Walsh. 2013. Human rhinoviruses. Clin. Microbiol. Rev. 26(1): 135–162.

Jernigan, J.A., B.S. Lowry, F.G. Hayden, S.A. Kyger, B.P. Conway, D.H. Groschel et al. 1993. Adenovirus type 8 epidemic keratoconjunctivitis in an eye clinic: risk factors and control. J. Infect. Dis. 167(6): 1307–1313.

Job, C.K. 1990. Nasal mucosa and abraded skin are the two routes of entry of Mycobacterium leprae. Star. 49(1).

Johnson, M.C., J.J. Bozzola, I.L. Shechmeister and I.L. Shklair. 1977. Biochemical study of the relationship of extracellular glucan to adherence and cariogenicity in *Streptococcus mutans* and an extracellular polysaccharide mutant. J. Bacteriol. 129(1): 351–357.

Katragadda, L., Z. Shahid, A. Restrepo, J. Muzaffar, D. Alapat and E. Anaissie. 2013. Preemptive intravenous immunoglobulin allows safe and timely administration of antineoplastic therapies in patients with multiple myeloma and parvovirus B19 disease. Transplant Infectious Disease 15(4): 354–360.

Kazda, J., R. Ganapati, C. Revankar, T.M. Buchanan, D.B. Young and L.M. Irgens. 1986. Isolation of environment-derived Mycobacterium leprae from soil in Bombay. Lepr. Rev. 57 Suppl 3: 201–208.

Keeffe, E.B. 2006. Hepatitis A and B superimposed on chronic liver disease: vaccine-preventable diseases. Trans. Am. Clin. Climatol. Assoc. 117: 227–37; discussion 237–238.

Kelly, H.A., D. Siebert, R. Hammond, J. Leydon, P. Kiely and W. Maskill. 2000. The age-specific prevalence of human parvovirus immunity in Victoria, Australia compared with other parts of the world. Epidemiol. Infect. 124(3): 449–457.

Khan, M.Y., M.W. Mah and Z.A. Memish. 2001. Brucellosis in pregnant women. Clin. Infect. Dis. 32(8): 1172–1177.

Koff, R.S. 1998. Hepatitis A. Lancet 351(9116): 1643–1649.

Krech, U. 1973. Complement-fixing antibodies against cytomegalovirus in different parts of the world. Bull World Health Organ 49(1): 103–106.

Krohn, M.A., S.L. Hillier and C.J. Baker. 1999. Maternal peripartum complications associated with vaginal group B streptococci colonization. J. Infect. Dis. 179(6): 1410–1415.

Kwartler, J.A. and A. Limaye. 1989. Pathologic quiz case 1. Cervicofacial actinomycosis. Arch. Otolaryngol. Head Neck Surg. 115(4): 524–527.

Lastoria, J.C. and M.A. Abreu. 2014. Leprosy: review of the epidemiological, clinical, and etiopathogenic aspects—part 1. An Bras Dermatol. 89(2): 205–218.

Liaw, Y.F., S.L. Tsai, I.S. Sheen, M. Chao, C.T. Yeh, S.Y. Hsieh et al. 1998. Clinical and virological course of chronic hepatitis B virus infection with hepatitis C and D virus markers. Am. J. Gastroenterol. 93(3): 354–359.

Liefeldt, L., M. Buhl, B. Schweickert, E. Engelmann, O. Sezer, P. Laschinski et al. 2002. Eradication of parvovirus B19 infection after renal transplantation requires reduction of immunosuppression and high-dose immunoglobulin therapy. Nephrol. Dial. Transplant. 17(10): 1840–1842.

Liu, A.Y., C.S. Nabel, B.S. Finkelman, J.R. Ruth, R. Kurzrock, F. van Rhee et al. 2016. Idiopathic multicentric Castleman's disease: a systematic literature review. Lancet Haematol. 3(4): e163–175.

Loiez, C., F. Wallet, P. Pischedda, E. Renaux, E. Senneville, N. Mehdi et al. 2007. First case of osteomyelitis caused by "Staphylococcus pettenkoferi". J. Clin. Microbiol. 45(3): 1069–1071.

Maartens, G., C. Celum and S.R. Lewin. 2014. HIV infection: epidemiology, pathogenesis, treatment, and prevention. Lancet 384(9939): 258–271.

Mahajan, S.K., J.M. Rolain, N. Sankhyan, R.K. Kaushal and D. Raoult. 2008. Pediatric scrub typhus in Indian Himalayas. Indian J. Pediatr. 75(9): 947–949.

Mammina, C., C. Bonura, M.S. Verde, T. Fasciana and D.M. Palma. 2011. A fatal bloodstream infection by *Staphylococcus pettenkoferi* in an Intensive Care Unit Patient. Case Rep. Crit. Care 2011: 612732.

Martins, A.C., A. Miranda, M.L. Oliveira, S. Buhrer-Sekula and A. Martinez. 2010. Nasal mucosa study of leprosy contacts with positive serology for the phenolic glycolipid 1 antigen. Braz J. Otorhinolaryngol. 76(5): 579–587.

Melsom, R., M. Harboe, M.E. Duncan and H. Bergsvik. 1981. IgA and IgM antibodies against Mycobacterium leprae in cord sera and in patients with leprosy: an indicator of intrauterine infection in leprosy. Scand. J. Immunol. 14(4): 343–352.

Munoz, N., F.X. Bosch, S. de Sanjose, R. Herrero, X. Castellsague, K.V. Shah et al. 2003. Epidemiologic classification of human papillomavirus types associated with cervical cancer. N Engl. J. Med. 348(6): 518–527.

Narvencar, K.P., S. Rodrigues, R.P. Nevrekar, L. Dias, A. Dias, M. Vaz et al. 2012. Scrub typhus in patients reporting with acute febrile illness at a tertiary health care institution in Goa. Indian J. Med. Res. 136(6): 1020–1024.

Nemerow, G.R., C. Mold, V.K. Schwend, V. Tollefson and N.R. Cooper. 1987. Identification of gp350 as the viral glycoprotein mediating attachment of Epstein-Barr virus (EBV) to the EBV/C3d receptor of B cells: sequence homology of gp350 and C3 complement fragment C3d. J. Virol. 61(5): 1416–1420.

Pappas, G., P. Panagopoulou, L. Christou and N. Akritidis. 2006. Brucella as a biological weapon. Cell Mol. Life Sci. 63(19-20): 2229–2236.

Parkman, P.D. 1996. Togaviruses: Rubella Virus. *In*: Baron, S. (ed.). Medical Microbiology. Galveston (TX).

Pedley, J.C. 1967. The presence of M. leprae in human milk. Lepr. Rev. 38(4): 239–242.

Pfaller, M.A. and L.A. Herwaldt. 1988. Laboratory, clinical, and epidemiological aspects of coagulase-negative staphylococci. Clin. Microbiol. Rev. 1(3): 281–299.

Phares, C.R., R. Lynfield, M.M. Farley, J. Mohle-Boetani, L.H. Harrison, S. Petit et al. 2008. Epidemiology of invasive group B streptococcal disease in the United States, 1999–2005. JAMA 299(17): 2056–2065.

Regan, J.A., M.A. Klebanoff, R.P. Nugent, D.A. Eschenbach, W.C. Blackwelder, Y. Lou et al. 1996. Colonization with group B streptococci in pregnancy and adverse outcome. VIP Study Group. Am. J. Obstet. Gynecol. 174(4): 1354–1360.

Robertson, S.E., D.A. Featherstone, M. Gacic-Dobo and B.S. Hersh. 2003. Rubella and congenital rubella syndrome: global update. Rev. Panam Salud Publica/Pan Am. J. Public Health 14(5): 306–315.

Rovery, C., A. Etienne, C. Foucault, P. Berger and P. Brouqui. 2005. Veillonella montpellierensis endocarditis. Emerg. Infect. Dis. 11(7): 1112–1114.

Salyers, A.A. 1984. Bacteroides of the human lower intestinal tract. Annu. Rev. Microbiol. 38: 293–313.

Santos, A.R., V. Balassiano, M.L. Oliveira, M.A. Periera, P.B. Santos, W.M. Degrave et al. 2001. Detection of Mycobacterium leprae DNA by polymerase chain reaction in the blood of individuals, eight years after completion of anti-leprosy therapy. Mem. Inst. Oswaldo Cruz 96(8): 1129–1133.

Schaberg, D.R., D.H. Culver and R.P. Gaynes. 1991. Major trends in the microbial etiology of nosocomial infection. Am. J. Med. 91(3B): 72S–75S.

Schrag, S., R. Gorwitz, K. Fultz-Butts and A. Schuchat. 2002. Prevention of perinatal group B streptococcal disease. Revised guidelines from CDC. MMWR Recomm. Rep. 51(RR-11): 1–22.

Sharp, P.M. and B.H. Hahn. 2011. Origins of HIV and the AIDS pandemic. Cold Spring Harb Perspect. Med. 1(1): a006841.

Shepard, C.C. 1962. The nasal excretion of Mycobacterium leprae in leprosy. Int. J. Lepr. 30: 10–18.

Siegel, J.D., E. Rhinehart, M. Jackson, L. Chiarello and Committee Health Care Infection Control Practices Advisory. 2007. 2007 Guideline for isolation precautions: Preventing transmission of infectious agents in health care settings. Am. J. Infect. Control. 35(10 Suppl 2): S165–164.

Sievert, D.M., P. Ricks, J.R. Edwards, A. Schneider, J. Patel, A. Srinivasan et al. 2013. Antimicrobial-resistant pathogens associated with healthcare-associated infections: summary of data reported to the National Healthcare Safety Network at the Centers for Disease Control and Prevention, 2009–2010. Infect. Control Hosp. Epidemiol. 34(1): 1–14.

Singh, A.E. and B. Romanowski. 1999. Syphilis: review with emphasis on clinical, epidemiologic, and some biologic features. Clin. Microbiol. Rev. 12(2): 187–209.

Sommer, A. and K.P. West. 1996. Vitamin A Deficiency. Health, Survival, and Vision. New York: Oxford University Press.

Trulzsch, K., H. Rinder, J. Trcek, L. Bader, U. Wilhelm and J. Heesemann. 2002. "Staphylococcus pettenkoferi", a novel staphylococcal species isolated from clinical specimens. Diagn. Microbiol. Infect. Dis. 43(3): 175–182.

Vaneechoutte, M., G. Verschraegen, G. Claeys, B. Weise and A.M. Van den Abeele. 1990. Respiratory tract carrier rates of Moraxella (Branhamella) catarrhalis in adults and children and interpretation of the isolation of M. catarrhalis from sputum. J. Clin. Microbiol. 28(12): 2674–2680.

Vivekanandan, M., A. Mani, Y.S. Priya, A.P. Singh, S. Jayakumar and S. Purty. 2010. Outbreak of scrub typhus in Pondicherry. J. Assoc. Physicians India 58: 24–28.

Walsh, G.P., W.M. Meyers, C.H. Binford, P.J. Gerone, R.H. Wolf and J.R. Leininger. 1981. Leprosy—a zoonosis. Lepr. Rev. 52 Suppl 1: 77–83.

Wang, Q., S. Fidalgo, B.J. Chang, B.J. Mee and T.V. Riley. 2002. The detection and recovery of Erysipelothrix spp. in meat and abattoir samples in Western Australia. J. Appl. Microbiol. 92(5): 844–850.

Wasley, A., A. Fiore and B.P. Bell. 2006. Hepatitis A in the era of vaccination. Epidemiol. Rev. 28: 101–110.

Wexler, H.M. 2007. Bacteroides: the good, the bad, and the nitty-gritty. Clin. Microbiol. Rev. 20(4): 593–621.

White, N.J., S. Pukrittayakamee, T.T. Hien, M.A. Faiz, O.A. Mokuolu and A.M. Dondorp. 2014. Malaria. Lancet 383(9918): 723–735.

Zaleznik, D.F., M.A. Rench, S. Hillier, M.A. Krohn, R. Platt, M.L. Lee et al. 2000. Invasive disease due to group B Streptococcus in pregnant women and neonates from diverse population groups. Clin. Infect. Dis. 30(2): 276–281.

CHAPTER 3

Microbiological Diagnosis of Bacterial Diseases

Sarita Mohapatra,[1,*] *Arti Kapil*[2] and *Nancy Khardori*[3]

INTRODUCTION

Infections due to common bacteria remain a major cause of morbidity and mortality among humans in the entire world. A recent report from World Health Organization (WHO) highlights bacterial infection as the major killer of children and young adults globally. The ability to detect the etiological agent correctly needs the combination of good clinical practice in terms of in-depth knowledge regarding epidemiology, host-parasite interaction, pharmacokinetics of the antibiotics and clinical microbiology laboratory facilities. Accurate diagnosis of bacterial diseases is essential not only for prompt and appropriate management but also to avoid the misuse/overuse or abuse of antibiotics.

The pathogenesis of bacterial infection is a combination of pathology due to mere presence of the bacteria and the consequences of inflammation caused by them. Hence, more than one test is usually needed for confirmation or exclusion of diagnosis. The primary requisite for establishing a correct diagnosis is appropriate sampling, transport/storage and processing of the specimen in the laboratory. The diagnostic cycle of laboratory assessment of any clinical specimen is divided into three categories (Fig. 1).

[1] Assistant Professor, Department of Microbiology, Bacteriology Lab, All India Institute of Medical Sciences, New Delhi, India-110029.

[2] Professor, Department of Microbiology, All India Institute of Medical Sciences, New Delhi, India.

[3] Professor of Medicine and Microbiology and Molecular Cell Biology, Director, Division of Infectious Diseases, Department of Internal Medicine, Eastern Virginia Medical School, Norfolk, Virginia.

* Corresponding author

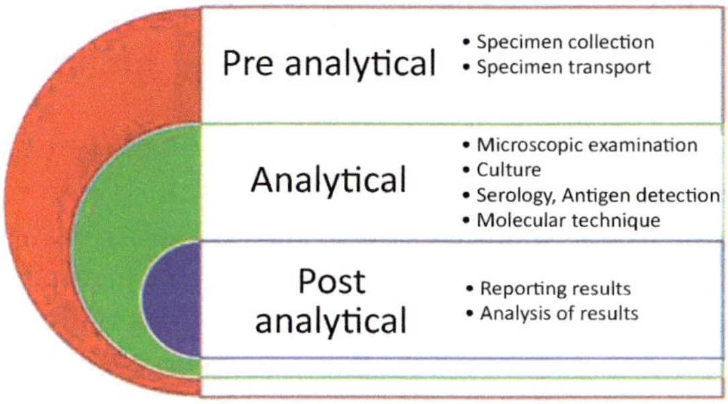

Fig. 1. Diagnostic Cycle.

All the three phases are equally important. In this chapter we first discuss the pre-analytical phase, followed by the diagnostic or analytical section. The analytical section has been described based on various anatomical sites such as blood stream infection, respiratory tract infection, central nervous system infection, genito-urinary tract infection, bone and joint infection, skin and soft tissue infection, etc.

Pre-analytical Phase

Pre analytical phase includes the correct identification of the pathogen based on clinical manifestation, appropriate sampling, and transport of the specimen within appropriate time (Baron 2013a). Appropriate sampling is the most important component amongst all.

A. Specimen collection (Baron 2013a, Collee 2014, Hassan 2014)

i. Specimen should be collected from the actual site of infection with minimum contamination from the adjacent site. For example
   Collection of sputum sample with minimum contamination with oropharyngeal secretion.
   Collection of throat swab from tonsillar fossa for streptococcal pharyngitis without contact with the oral flora.
   Collection of sample from abscess or sinuses. Failure to reach the depth of abscesses or sinuses may give a false negative report. Surface culture, on the other hand may give a false positive result.
ii. Optimal time of collection for optimal recovery of the causative agent: Prior knowledge regarding the pathophysiology of the condition is important for determining optimal recovery of the causative microorganism. For Example:

Blood for culture in typhoid fever should be collected in the first week of disease serological testing should be done during the second week of presentation.

iii. Sufficient quantity of specimen: Too little quantity of sample or dry swab may give a false negative culture result. When the sample size is very small priority should be given to culture rather than Gram Staining. To prevent drying of swabs and for optimal recovery of the microbiological agent, tube with holding media such as normal saline or phosphate yeast glucose broth or transport media, e.g., Amie's/ Stuart's media should be used.

iv. Appropriate specimen collection, Containers and Culture media. The containers should be sterile and designed for ease of collection. They should be properly labeled with at least patient's name and identification number, source of specimen, date/time of collection.
Wide mouthed, screw capped, leak proof containers are commonly used for the collection of samples such as sputum, urine, stool.
Swabs are generally inferior to other methods of specimen collection. The use of swabs in anaerobic infection should be discouraged. Cotton swabs may contain fatty acids, which inhibit the bacterial growth. Similarly, Calcium alginate swabs may release toxic products, which further hamper the growth of the bacteria. Dacron, Rayon or Polyester swabs should be used for better recovery of fastidious organisms. Ability to absorb and release specimen is another important criteria for the selection of swabs. This varies from material to material. Currently, commercially available flocked swabs are preferred over conventional swabs. The brush bristles of flocked swab are better at retaining the specimen. They also possess numerous microscopic folds, which significantly increase the surface and also allow release of specimen more efficiently for further testing.
The specimen for culture should always be obtained before the administration of antibiotics

B. Specimen transport

The primary objective of the specimen transport is to maintain and recover the microbiological agent in its original form. Most of the liquid specimens are recommended to reach to the laboratory within 2 hrs of collection. However, urine samples may be refrigerated overnight or can be transported in the containers containing 1.8% boric acid. A number of holding media or transport media can be chosen depending upon the type of specimen, the detail will be discussed under analytical section.

Analytical Phase

Analytical phase is the phase of diagnosis, which deals with the processing of the specimen in the clinical microbiology laboratory.

A. Infection of blood and cardiovascular system

Blood is the most important sterile fluid in the body. Invasion of microorganism into blood has serious consequences. Early and correct diagnosis of blood stream infection (BSI) significantly decreases the morbidity and mortality. Successful recovery of microorganism from blood depends upon many factors, e.g., type of infection, type of bacteremia (shedding of microorganism in the blood transiently/intermittently/continuously), number of specimens, volume of each specimen, time of collection of blood, patient population, and laboratory expertise.

The source of bacteria in BSI can be divided into intravascular or extravascular based on their origin. The examples of disease conditions leading to intravascular and extravascular BSI and common bacteria are given in Tables 1 and 2.

Table 1. Underlying Conditions Causing BSI.

Intravascular causes	**Extravascular causes**
Infective endocarditis Mycotic aneurysm Intravascular device associated bacteraemia	Wound infection and soft tissues Infection of organ systems, e.g., Respiratory tract, Genitourinary tract, Reticuloendothelial system

Table 2. Bacteria Commonly Isolated from Blood Culture.

Agents of Infective endocarditis	**Agents of Vasular Catheter related-BSI**	**Agents commonly associated with BSI from extravascular sites**
Viridans Streptococci *Enterococcus* spp. *Staphylococcus aureus* Nutritionally deficient Streptococci HACEK group *Coagulase negative Staphylococci* *Enterobacteriaceae* *Haemophillus* spp. *Coxiella burnetii* Candida spp.	*Coagulase Negative Staphylococci* *Staphylococcus aureus* *Enterococcus* spp. *Pseudomonas aeruginosa* *Enterobateriaceae* *Corynebacterium jeikum* *Burkholderia cepacia complex* *Stenotrophomonas* spp. Candida spp.	*Anaerobic organism* *Coagulase negative Staphylococci* *Enterobacteriaceae* *Haemophillus influenzae* *Legionella* spp. *Listeria monocytogenes* *Neisseria meningitidis* *Pseudomonas aeriginosa* *Salmonella enterica typhi* *Streptococcus pneumoniae* *Streptococcus pyogenes* *Staphylococcus aureus* *Brucella* spp. *Chlamydia pneumoniae*

To detect BSI, blood sample must be collected aseptically and processed for cultures quickly. In any suspected primary site infection; samples such as CSF, respiratory samples, wound drainage, pus, urine, etc. should be collected and processed for cultures simultaneously.

Blood cultures

Collection of blood cultures with the objective of minimizing contamination involves the following steps.

1. Perform hand hygiene and wear sterile gloves before the procedure.
2. Wash the area with soap and water.
3. Apply povidone-iodine, tincture of iodine or chlorhexidine-gluconate and allow drying for 1–2 mins (povidone-iodine) or 30 secs (tincture of iodine/chlorhexidine-gluconate). Chlorhexidine-gluconate is not used in children less than 2 months old. Scrub the disinfectant starting from the center to the periphery encircling an area of 4 to 5 inches. Remove iodine from the skin by scrubbing with 70% alcohol.
4. Apply tourniquet and do not palpate the site with gloved fingers again.
5. Collect blood aseptically by needle and syringe or a closed vacutainer system.

The total blood volume being processed is much more important than the number of samples. Many studies have shown improvement in the yield of blood cultures with increase in the volume processed. Ideally 10–20 ml of blood in case of an adult and 5–10 ml in case of pediatric patients divided in two different bottles should be sent to the laboratory.

Primary site cultures as indicated clinically, e.g., CSF, respiratory secretion, wound discharge, pus and urine should be collected simultaneously.

The blood culture bottle should be immediately transferred to the laboratory. In case of delay, it can be kept in an incubator at 35°C or left at room temperature. But NEVER REFRIGERATE the bottle.

The blood can be processed either by manual or automated culture systems. Irrespective of the type of system, all the blood culture bottles should be incubated for at least five days before finalizing a negative report. In manual system, the bottle should be inspected daily for any turbidity, hemolysis, gas formation, pellicle formation, clot or visible colonies, which indirectly suggest the bacterial growth. In case of visually negative samples, blind subculture on the blood agar and MacConkey agar plates should be performed after overnight incubation and on 5th day of incubation. One can also perform gram stain, motility or acridine orange stain from the culture bottle to confirm the growth of bacteria.

Automated and computerized culture systems are currently the preferred standard for blood culture in most of the clinical microbiology laboratories. The significant advantages of these systems are the continuous reading, automated detection and computerized technique; which decrease time to positivity and reporting in comparison to manual blood culture systems. Three such systems are currently available: BacT/ALRT 3D (bioMerieux, Durham, NC), BACTEC (BD Diagnostics Systems), Versa TREK (ThermoScientific, Cleveland, OH). Each system gives signal as soon as the growth is detectable and alerts the microbiologist to further confirmation by gram stain and/or subculture.

Special consideration should be given to the potential of fastidious organisms and for patients with endocarditis; where the causative organisms may take longer period to grow. Such cultures are incubated for extended period. A discussion between the clinician and the laboratory technologist regarding the clinical suspicions helps with better recovery of pathogens in these instances.

Blood is a sterile specimen. Hence any growth of microbes from blood culture is an abnormal finding. However, one must consider that;

i. The skin microbiota may contaminate the sample during collection causing false positive result.
ii. Mixed culture can be significant in causation of disease. Hence, one should be vigilant while examining the subcultures on solid agar plates. Gram stain of each type of colony is helpful for further evaluation of mixed growth.
iii. One may get a false negative result if the number of microorganisms is low.
iv. Sodium-polyanethol sulphonate in blood culture bottles may inhibit the growth of some fastidious organisms.
v. Bacterial metabolism may not generate sufficient carbon dioxide to be detected by the automated systems.

Recently, FDA has approved multiplex molecular assays for rapid and accurate diagnosis from the signal positive blood culture bottles (Millar 2007). They not only identify the organism rapidly up to species level but also detect the genetic determinants of the antimicrobial resistance in select situations.

B. Infection of respiratory tract

Infection of the respiratory tract is another common infection where the role of clinical microbiology laboratory is very important. Anatomically, the respiratory tract is divided into upper and lower respiratory tract and microbiological diagnosis varies accordingly.

The upper respiratory tract (URT) consists of nose, throat, pharynx, epiglottis, and larynx. The common pathologic conditions are rhinitis, sinusitis, otitis media, pharyngitis, other oral cavity infections (e.g., gingivitis, Herpes simplex infection) epiglottitis, tonsillitis, cervical lymphadenitis, laryngitis, etc. The lower respiratory tract (LRT) consists of trachea, bronchi, and lungs. The infectious diseases are pneumonia, traheobronchitis, bronchitis, bronchiolitis, lung abscess, etc. The presenting symptoms vary depending on the age group and underlying conditions. The etiological agents and specimens for microbiological diagnosis in upper and lower respiratory tract infection are shown in Table 3.

The upper respiratory tract harbors a large number of commensal (normal flora) organisms. These organisms may have a role in the causation of disease whenever the host immunity is compromised or there is a breech in the anatomy. They play an important role in device-associated infections (endotracheal tube, tracheostomy tube). Moreover, they may contaminate the specimens from upper and lower respiratory tract during collection. Hence, clinicians must have sufficient knowledge about proper collection of specimens and interpretation of results.

Table 3. List of Etiological Agents and Specimens for Culture in Upper and Lower Respiratory Tract Infections.

Etiological agents	**Specimens for culture**
Streptococcus pneumoniae	**URI:**
β-*hemolytic streptococci*	Nasopharyngeal swab
Staphylococcus aureus	Sinus washings
Haemophillus influenzae	Swab of tonsils, swab from posterior pharyngeal wall
Klebsiella pneumoniae	
Other enterobactericeae	**LRI:**
Streptococcus spp., *Corynebacterium diptheriae*	Sputum
Bordetella pertussis	Bronchoalveolar secretion
Leigionella spp.	Transtracheal secretion
Yersinia pestis	Lung aspirate
Pseudomonas spp.	Biopsy
Acinetobacter spp.	Blood
Burkholderia spp.	
Mycobacterium tuberculosis	
Prevotella melanogenicus	
Fusobacterium nucleatum	

Throat and nasopharyngeal swabs: Commercially available sterile rayon or dacron swabs or flocked swabs are preferred to cotton swabs for the collection of specimens from upper respiratory tract. Calcium alginate or dacron swabs are especially recommended for nasopharyngeal swab specimens for diagnosis of bordetella infection.

Sputum: Sputum is usually formed due to infection in the lower respiratory tract infection. It is a mixture of bronchial secretions and inflammatory exudates during infection. Early morning collection of sputum sample is ideal as it contains pooled overnight secretions. Twenty-four hour sputum collection should be discouraged. Mouth gargles should not be advised just before the sample collection. The specimen should be collected in a wide mouthed, screw capped, leak proof, sterile container. The patient must be advised to direct sputum into the container after a bout of cough in order to minimize contamination by oropharyngeal flora or saliva. The specimen should reach the laboratory as soon as possible and processed within two hours of collection.

Endotracheal aspirate (EA): Endotracheal aspirate is collected by introducing a sterile catheter into the endotracheal tube. First, the collection inside the endotracheal tube is aspirated and discarded and then 5 ml of sterile normal saline is instilled into the tube and collected for processing. The specimen must be processed in the microbiology laboratory within one hour of collection.

Bronchoalveolar lavage (BAL): Bronchoalveolar lavage is preferred over EA for the diagnosis of lower respiratory tract infection. It is collected by instilling 100–300 ml of normal saline into pulmonary interstitium and alveolar space through a bronchoscope and retrieving a part of it for processing.

Other samples used for diagnosis of respiratory tract infection are translaryngeal/transtracheal aspiration, fine needle aspiration, lung biopsy, pleural fluid and pleural biopsy.

Rejection criteria

Sputum and endotracheal aspirate should not be processed if

- Collected for 24 hours prior to the receiving in the laboratory
- Duplicate sample on same day
- Repeated within 48 hours

Sputum specimen can be rejected if

- There is gross salivary contamination
- Microscopic examination shows more than 10 epithelial cell/oil immersion field
- Sample was collected after mouthwash or gargle

BAL, lung biopsy and fine needle aspirate and other samples collected after invasive procedures should always be processed.

Processing and interpretation

All the samples should undergo microscopy (gram strain) and culture except EA. Gram stain; Albert's stain is useful in throat swabs for detection of *Corynebacterium diptheriae* infection. Microscopy plays an important role in determination of appropriateness of the sample in case of lower respiratory tract specimens such as sputum. Presence of white blood cells in BAL samples indicates inflammation and further correlates with the culture findings. The specimens can be cultured on 5% sheep blood agar, chocolate agar and MacConkey agar. Chocolate agar is used to differentiate Haemophillus from *Neisseria* spp. The cultures of EA and BAL specimens are interpreted semi quantitatively. Colony counts $\geq 10^5$/ml in EA and $\geq 10^4$/ml in BAL samples are considered significant. Special media like Buffered charcoal yeast extract (BCYE) can be used in suspected cases of Legionella infection. Often blood cultures are reliable and useful for managing lower respiratory tract infections.

C. *Infection of urinary tract*

Urine is the most commonly used specimen for detection of urinary tract infection (UTI). Sometimes the ascending infections may manifest as bacteremia secondary to UTI. Hence, blood samples are also taken in addition to urine to establish the diagnosis. Urinary tract in women is lined by large number of commensals. There is high probability of contamination of the sample with the commensals of urethra and the perineal flora during the process of collection. Urine is a very good media to allow the growth of bacteria. Hence, proper sampling is very crucial.

Urine cultures

Clean catch midstream urine is the most common method for collection of urine specimens. Urine can also be collected from a patient with devices like Foley's catheter, nephrostomy tube, cystoscope, suprapubic aspiration, illeal conduit and by tapping method for neonates.

Clean catch midstream

The urine sample should be collected in wide mouthed, screw capped, leak proof containers.

- Use of antiseptic solutions should be discouraged. The patients should be advised to wash the perineal area with soap and water to minimize the contamination rate.
- The first morning voided urine sample should be collected as it is expected to contain maximum number of bacteria.
- The first portion of the urine sample should be discarded to decrease contamination with the urethral commensals.

Indwelling catheter and nephrostomy tube

- Urine MUST NOT BE obtained from the collection bag in a catheterized patient.
- A clamp should be applied on the portion of the catheter below the junction. The area of the catheter chosen for collection should be disinfected with 70% alcohol. Using a sterile syringe urine should be aspirated after waiting for 10 minutes to collect a fresh specimen.

The urine sample should be transported to the laboratory as quickly as possible and processed within two hours. In case of delay urine can be refrigerated up to 24 hours. Suprapubically aspirated specimen should not be rejected as it is collected by an invasive procedure.

Wet mount preparation of the uncentrifuged urine specimen can be examined under microscope for the presence of leukocytes, RBCs, epithelial cells, crystals, casts, and organisms. Several screening tests like nitrate reduction test, catalase test, leukocyte reductase/esterase test are also available which can indirectly indicate presence of bacterial or polymorphonuclear cell enzymes. But these tests are less specific, and only good for screening. The urine is cultured on Cysteine Lactose Electrolyte Deficient (CLED) media. Recently a chromogenic medium for urinary tract pathogens named BD CHROMagar (Becton Dickison, Heidelberg, Germany) has become available for the presumptive diagnosis of uropathogens directly from the urine specimen. The final interpretation of urine culture is done by counting the number of colony forming units (CFU) on the agar plate as shown in Table 4.

Table 4. Interpretation of Urine Culture based on CFU/ml.

CFU/ml	Significance
$\geq 10^5$/ml	Significant bacteriuria, UTI highly likely
$> 10^3$–$\leq 10^5$/ml	Doubtful significance, probable UTI
$\leq 10^3$/ml	Insignificant bacteriuria, UTI unlikely (EXCEPT samples collected by Suprapubic aspiration, percutaneous nephrostomy, cystoscopy)

Significant colony count of 1–2 different organisms in the presence of WBC should be reported as significant bacteriuria and antimicrobial susceptibility testing should be done either for the predominant morphotype or both depending upon the clinical factors. Presence of three or more types of organisms shows gross contamination during collection. Hence, results on such specimens should be reported as contamination or mixture of organisms and a repeat sample with proper collection should be advised to evaluate the clinical significance.

D. Infection of gastrointestinal system

The infection of gastrointestinal tract occurs from ingestion of contaminated food and water. These infections may be associated with diarrhea with or without abdominal pain or vomiting. The various microorganisms responsible for the causation of gastrointestinal infectious diseases are given below in Table 5. The microorganisms cause GI diseases either due to toxin production or invasion causing mucosal inflammation (Procop 2017).

Table 5. Bacteria causing Gastrointestinal Infections.

Mechanisms	Bacterial etiology
Enterotoxin	*S. aureus* *V. cholerae* *S. dysenteriae type 1* *Enterotoxin producing E. coli* *Salmonella* spp. *B. cereus* *Cl. difficile* *Aeromonas* spp. *C. jejuni*
Cytotoxin	*Shigella* spp. *Cl. difficile* (type B) Enterohemorrhagic *E. coli*
Neorotoxins	*Cl. botulinum* *B. cereus*
Invasion	*Shigella* spp. *Enteroinvasive E. coli* *C. jejuni* *Y. enterocolitica*
Attachment and effacement	Enteropathogenic *E. coli*

Stool culture

Feces is the most common specimen collected for diagnosis. Rectal swab is not satisfactory, hence should be discouraged. If food poisoning is suspected, food particles and vomitus can be collected for processing. The stool sample should be collected in a wide mouthed, screw capped, leak proof container. The sample should reach the laboratory without delay. In case of delay it should be sent in transport media like buffered glycerol saline, Stuart's transport medium, Amie's transport medium, etc. The stool sample should be first examined visually followed by microscopy and culture. The sample should be inspected for presence of mucus, pus and blood. Presence of blood is indicative of dysentery. One should always examine a hanging drop immediately irrespective of clinician request. Darting motility under microscope should be immediately reported to the clinician as presumed diagnosis of cholera. Stool sample should be cultured on blood agar, MacConkey agar, Deoxycholate citrate agar, Xylose lysine deoxycholate agar and selenite F broth, etc. Specimens suggestive of cholera infection should be inoculated in thiosulphate citrate bile salt sucrose (TCBS) agar.

E. *Infection of central nervous system and other body sites with sterile specimens*

Infection of central nervous system is a life-threatening condition and needs urgent evaluation and treatment. Microbiology laboratory plays an important role not only in establishing the etiological agent but also in helping start appropriate antibiotic treatment. Most common infections are meningitis, encephalitis, meningoencephalitis, myelitis and brain abscess. Cerebrospinal fluid (CSF) specimen is the most reliable specimen for laboratory investigation of CNS infection. The etiological agents are different for different age groups of patients and also depend on a number of host factors. Both the clinician and microbiologist should be aware of these factors so that appropriate processing is done in the microbiology laboratory. The association of etiological agents in different groups of patients is given in Table 6.

Table 6. Distribution of Etiological Agents in Different Group of Patients.

Host	**Etiological agents**
Neonates (< 1 month)	*Escherichia coli, Gr. B Streptococci*
≥ 1 month–< 2 month	*Gr. B Streptococci, Haemophillus influenzae*
≥ 2 month–< 10 year	*Haemophillus influenzae, Streptococcus pneumoniae*
Adults	*Streptococcus pneumoniae, Neisseria meningitidis*
Immunocompromised	*Cryptococcus neoformans, Listeria monocytogenes*
Patients with shunts	CoNS

CSF cultures

The usual specimen obtained in case of central nervous system infection is CSF. It should be collected by a trained clinician with optimum aseptic precautions. After collection it should be submitted to the laboratory in three parts:

Tube 1: CSF for cell count

Tube 2: CSF for gram's stain and culture

Tube 3: CSF for biochemical test such as glucose and protein, and other special test such as, VDRL, Cryptococcal antigen test, etc.

Tube 4: CSF reserved for further testing depending on the results of above.

After collection, CSF should be immediately transferred to microbiology laboratory for processing. In case of delay it can be kept at room temperature overnight BUT SHOULD NEVER BE REFRIGERATED. Assessment of cell count and biochemical tests is very crucial for early guidance towards treatment (Table 7).

Table 7. Different Parameters in the Cerebrospinal Fluid in Meningitis.

Etiology	Cell Type	Glucose	Protein
Bacteria	100–1000/mm^3 Neutrophils	Low	High
Mycobacteria	100–500/mm^3 Lymphocytes	Low	High
Viruses	1–500/mm^3 (Lymphocytes)	Normal	Normal

Once the CSF specimen arrives in the laboratory, its gross appearance should be examined. Wet mount of the centrifuged CSF sample should be immediately viewed under compound microscope. The microbiologist must report the number of WBCs, RBC, bacteria and yeast cells. Dark ground microscope should be used in suspected cases of leptospirosis. After the cell count the CSF should be centrifuged to get deposits of cells and bacteria. From the deposit a smear should be made for gram stain and culture should be done with aseptic precautions. Since the number of bacteria in the CSF is low, one should make a heaped up smear without spreading followed by air drying. The smear should be fixed with heat or alcohol. Any bacteria in the gram-stain are considered significant. The report of gram stain of CSF should be made immediately available to the clinician. Simultaneously CSF and serum should also be tested for the bacterial antigens and cryptococcal antigen. The deposits should be cultured on the solid agar plates like blood agar, MacConkey agar and other special agar plates in case of fastidious organisms. Scarping from the petechial rash can show organisms on gram stain in meningococcal disease.

Sterile fluids from other body sites

Pericadial fluid, pleural fluid, amniotic fluid, synovial fluids are examples of sterile body fluids. The most important thing to be remembered is the collection of the sample with maximum precaution to avoid contamination with the skin flora. All the fluids should be tested by microscopy and plated on blood agar and MacConkey agar for the microbiological diagnosis.

F. Wound infections, abscesses, ulcers and soft tissue infections (Baron 2013b)

Infection of the any wound or intact soft tissue is caused mostly by bacteria. These bacteria may originate from exogenous source or endogenous sources. The superficial wounds are usually colonized by the skin commensals. Therefore, it becomes difficult to define and distinguish pathogens from colonization or contamination. Superficial wounds should be irrigated with normal saline and closed wound should be disinfected with 70% alcohol or chlorhexidine before collection of samples. Wounds should be curetted to clean the superficial deposits (biofilms) and biopsy or scraping from the base of the lesion should be obtained for processing. Collection of pus sample or exudate on swabs should be discouraged unless used as a part of a system which maintains moisture and viability. Ideally two swabs should be taken: one for microscopy, other for culture. Swab specimens should be processed in the laboratory without delay. In case of delay the swabs should be transported in tubes containing transport media and kept at room temperature but not refrigerated. If the swab has dried, it should be moistened with normal saline before making smear in the laboratory.

Naked eye examination is the initial step to comment on any pigmentation/smell (fruity/fishy)/necrosis/sulphur granules. If only one swab is available, one should always go for culture followed by gram stain to avoid contaminats in culture. The specimens should be inoculated on blood agar and MacConkey agar.

Post-analytical Phase

The post analytic part includes the correct interpretation and reporting of results. This includes examination of culture for colony morphology, identification of bacteria by staining, and biochemical tests. Along with the conventional techniques, several screening tests based on chromogenic characteristics [RapID SS/u System (Remel), NGP-Wee-Tabs (KEY Scientific Products, Round Rock, TX)] are available for rapid detection of *Enterobacteriaceae*. Chromogenic agar media containing various substrates such as CHROMagar Orientation (BD Diagnostics, Sparks, MD, Hardy Diagnostics, Santa Maria, CA) and Chromogenic UTI (Oxford, Basingstoke,

United Kingdom) are also available. They are able to differentiate various genera of bacteria based on appearance of the colonies. This occurs due to the presence of specific microbial enzymes; which produce colored compounds by hydrolyzing the substrate present in the media.

Several semi automated and automated systems named MicroScan Walkway (Beckman Coulter, West Sacramento, CA), Vitek System (BioMerioux), Sensititer Automated Reading and Incubation System (ARIS) (Thermo Fisher Scientific Inc., Waltham, MA), Phoenix system (Becton Dickinson Microbiology Systems, Sparks, MD) are now available for bacterial identification. Vitek and Phoenix systems can also perform antimicrobial susceptibility testing along with identification. Recent use of Mass spectrometry has significantly decreased the time for bacterial identification up to species level. This equipment is FDA approved and based on matrix associated laser desorption time of flight (MALDI-TOF) principle. The fresh culture isolate is put on the matrix and allowed to be hit by the laser. The protein biomolecules are ionized by the laser and subsequently separated from each other. The time of flight of each biomolecule is recorded and signal of all molecules forms a signature for the identification of that particular organism. This gives results within minutes, that are, reliable and reproducible.

In vitro susceptibility testing

Finally, the role of the microbiologist is to test the identified bacteria for the antimicrobial susceptibility.

Summary

In summary, preliminary reports should be generated when there is delay in identification or growth. An interim report on negative result is also helpful for the clinician to rethink the sampling and/or de-escalation of antimicrobials, etc. Antimicrobial stewardship is the key to prevent the antimicrobial resistance, and other complications like *Clostridium difficile* diarrhea, antimicrobial toxicity, adverse drug reaction, etc. The six Ds of antibiotic stewardship program, which emphasize the role of microbiology laboratory are—"Diagnosis, Drainage/Debridement, Drug, Dose, Duration, and De-escalation. Microbiology data helps administer appropriate antibiotic therapy; thereby minimizing antibiotic resistance. It is also proven to be beneficial in improving patient outcomes by rapid administration of the appropriate antibiotic dose, de-escalation or escalation of antibiotics in critical patients, reducing hospital acquired infections, and saving the hospital cost. Use of right dose according to the diagnosis, site of infection, renal or hepatic dysfunction is very important. Skin commensals should not be reported as pathogens in cases of non-critical patients except in clinically relevant samples such as blood or prosthetic joints. Microbiology data also give further clinical guidance in conditions with polymicrobial infection, diagnostic follow-up, e.g., repeat blood culture in candidaemia, link to respiratory viruses in lower respiratory tract infections, etc. Clinical microbiologist along with infectious disease specialist should collaborate for proper implementation of antibiotic stewardship program in their day-to-day practice. Biological markers like interleukins, Procalcitonin; C-reactive proteins, etc. are also used as an adjunct to the diagnosis of bacterial diseases. High Procalcitonin level especially has been seen associated with bacterial sepsis. Molecular techniques like PCR, nested PCR, nucleic acid hybridization, RNA typing, ribotyping, whole genome sequencing can also be done for bacterial identification. PCR is useful for the detection of antimicrobial resistance genes like extended spectrum beta-lactamases, carbapenemase, mecA gene for methicillin resistance, vancomycin resistance genes, etc. Clinical microbiology is an integral part of managing infectious diseases. Clinician should use the clinical symptoms/signs and initial laboratory investigation for early and correct identification of the disease process and collection of appropriate specimens. An appropriate presumptive antibiotic regimen can be started based on common pathogens responsible for infections of the organ system involved. The local, regional and national data on antibiotic resistance patterns should serve as a guide.

Suggested Readings

Baron, E.J., J.M. Miller, M.P. Weinstein, S.S. Richter, P.H. Gilligan, R.B. Thomson, Jr. et al. 2013a. Executive summary: A guide to utilization of the microbiology laboratory for diagnosis of Infectious diseases: 2013 recommendations by the Infectious disease Society of America (IDSA) and the American Society of Microbiology (ASM) (a). Clin. Infect. Dis. 57: 485–8.

Baron, E.J., J.M. Miller, M.P. Weinstein, S.S. Richter, P.H. Gilligan, R.B. Thomson, Jr. et al. 2013b. A guide to utilization of the microbiology laboratory for diagnosis of Infectious diseases: 2013 recommendations by the Infectious Disease Society of America (IDSA) and the American Society of Microbiology (ASM). Clin. Infect. Dis. 57: e22–e121.

Collee, F.G., F.P. Duguid, A.G. Fraser, B.P. Marmion and A. Simmons. 2014. Laboratory strategy in the diagnosis of infective syndromes. pp. 53–94. *In*: Collee, J.G., A.G. Fraser, B.P. Marmion and A. Simmons (eds.). Mackie & McCartney Practical Medical Microbiology (14th ed.). Churchill Livingstone; Elsevier. New Delhi.

Hassan A. Aziz, Maribeth L. Flaws, Lynne S. Garcia, Laurie A. Gregg, April L. Harkins and Rita M. Heuertz. 2014. Diagnosis by organ system. pp. 860–981. *In*: Tille, Patricia M. (ed.). Bailey & Scott's Diagnostic Microbiology (13th ed.). Elsevier, China.

Millar, B.C., Z. Xu and J.E. Moore. 2007. Molecular diagnostics of medically important bacterial infection. Curr. Issues Mol. Biol. 9: 21–39.

Procop, Gary W., Deirdre L. Church, Geraldine S. Hall, William M. Janda, Elmer W. Koneman, Paul C. Schreckenberger et al. 2017a. Introduction to microbiology Part II guidelines for the collection, transport, processing, analysis and reporting of cultures from the specific specimen sources. pp. 66–110. *In*: Washington, C.J. Jr., Stephen D. Allen and Geraldine S. Hall (eds.). Koneman's Color Atlas and Textbook of Diagnostic Microbiology. Woltes Kluwer, China.

Procop, Gary W., Deirdre L. Church, Geraldine S. Hall, William M. Janda, Elmer W. Koneman, Paul C. Schreckenberger et al. 2017b. Enterobacteriaceae. pp. 214–302. *In*: Washington, C.J. Jr., Stephen D. Allen and Geraldine S. Hall (eds.). Koneman's Color Atlas and Textbook of Diagnostic Microbiology. Woltes Kluwer, China.

CHAPTER 4

The Gram Stain Revisited: The Original Rapid Diagnostic Test

Role in Antibiotic Selection and Antibiotic Stewardship Programs (ASPs)

Cheston B. Cunha[1] and *Burke A. Cunha*[2,*]

INTRODUCTION

In this era of rapid diagnostic methods, there is renewed interest in the clinical usefulness of the Gram stain of body fluids to provide a presumptive microbiologic etiology upon which to guide initial antibiotic therapy. The Gram stain, the original "rapid diagnostic test", provides not only key information regarding the presumptive pathogen but also provides critical information on the associated cellular response, i.e., intensity of PMNs present. Only the Gram stain provides a rapid assessment of the pathogen and local host response in the body fluid being sampled. In antibiotic stewardship programs (ASP) the Gram stain provides a critical test to differentiate colonization from infection. One of the priorities in ASP is to identify infections (which should be treated) and to differentiate infection from colonization (which ordinarily should not be treated with antimicrobial therapy). In several important clinical scenarios Gram stain results interpreted in the light of clinical experience provides rapid and accurate differentiation between colonization and infection (Cunha and Cunha 2017, Cunha et al. 2013).

[1] Infectious Disease Division, Rhode Island Hospital and The Miriam Hospital; and Brown University Alpert School of Medicine, Providence, Rhode Island.

[2] Infectious Disease Division, Winthrop-University Hospital, Mineola, New York; and State University of New York, School of Medicine, Stony Brook, New York.

* Corresponding author: bacunha@winthrop.org

One of the aims of ASP is to optimize antimicrobial therapy. An important part of this objective is not to needlessly treat non-bacterial infections, e.g., viral infections with antibiotics. The cellular response component of the Gram stain assists in this determination. Another important ASP goal is to treat infection, but not colonization, e.g., asymptomatic bacteriuria (ASB). In ASP, carefully considered and selected antibiotic therapy will minimize the potential for antibiotic side effects and minimize subsequent antibiotic resistance (Cunha 2000, Hudepohl et al. 2016). Traditionally, the Gram stain of body fluids provides rapid preliminary identification of the likely bacterial pathogen in several important clinical setting, e.g., acute bacterial meningitis (ABM), community acquired pneumonia (CAP), urinary tract infections (UTIs), and wound infections (Table 1). The usefulness of the Gram stain in determining appropriate initial antibiotic therapy for wound infection cannot be overemphasized. A Gram stain smear of wound

Table 1. The Clinical Usefulness of the Gram Stain in Antibiotic Stewardship Programs (ASP).

- The Gram stain is underutilized and underappreciated and unequalled in speed and low cost. The original "rapid diagnostic method"
- Rapid microbiologic information is the basis for empiric antimicrobial therapy pending culture results
- The Gram stain has 2 key components (cellular response *plus* bacterial morphology/ arrangement)!

Colonization vs. Infection

- The Gram stain is the only laboratory test to practically/quickly differentiate colonization from infection:
- Colonization should not be treated
  - Pharyngitis
  - Wounds
  - Asymptomatic bacteriuria (ASB) or catheter associated bacteriuria (CAB)

Rapid Presumptive Diagnosis

Acute bacterial meningitis (AMB)

- Dx can be ruled out without significant CSF pleocytosis

Community acquired pneumonia (CAP)

- Dx Can be ruled out without significant sputum PMNs (in goal sputum specimen)
- Provides presumptive identification of CAP pathogens

Acute uncomplicated cystitis (AUC) and acute pyelonephritis (AP)

- Dx can be ruled out without significant pyuria and bacteriuria

Wound Infections

- Provides presumptive identification of wound pathogens

*clinical presentation: otherwise unexplained fever > 102 °F with CVA/flank tenderness.

Adapted from: Cunha, C.B. and B.A. Cunha. 2017. Antibiotic Essentials (15th ed.). JayPee Medical Publishers, New Delhi.

exudates, immediately differentiates between Gram positive cocci (GPC) in clusters (staphylococci) from GPC in chains (streptococci) permitting more selective antibiotic therapy. As with all laboratory tests, clinical relevance is dependent on interpretation in the appropriate clinical context, for example in wound exudates, GPC in clusters does not differentiate *S. aureus* from *S. epidermidis* (CoNS), and the Gram stain doesn't differentiate methicillin sensitive *Staphylococcus aureus* (MSSA) from methicillin resistant *Staphylococcus aureus* (MRSA). Clinically, although CoNS has the same Gram stain appearance as MSSA/MRSA, i.e., GPC in clusters, CoNS (unrelated to device associated infections) does not cause wound infections. For wound exudates with GPC in chains, MSSA/MRSA are immediately ruled out and microbiologically the differential diagnosis (DDx) is limited to streptococci, i.e., Group A streptococci (GAS) or Group B streptococci (GBS). In wounds, the Gram stain morphology (chain length) differentiates GAS/GBS (long chains) from enterococci (short chains), i.e., vancomycin resistant enterococci/vancomycin sensitive enterococci (VRE/VSE). Again, clinical correlation gives relevance to Gram stain DDx possibilities, i.e., VSE/VRE often colonize wounds, but are not sole wound pathogens. GPC in pairs, is indicative of *S. pneumoniae* and not streptococci or staphylococci. Once again, clinical correlation is useful since *S. pneumoniae* rarely, if ever, causes wound infections. With Gram negative bacilli (GNB) in wound exudates, the microbiologic DDx is narrowed by morphology, i.e., "plump" vs. "slender" GNB vs. usual size/shaped GNB. Plump/encapsulated GNB suggests *Klebsiella pneumoniae*, while slender/unencapsulated GNB suggests *P. aeruginosa*. The Gram stain cannot further differentiate other aerobic GNB in wounds (Bottone 2004, de la Maza et al. 2004, Cohen et al. 2016, Koneman et al. 1997, Madigan et al. 2003, Versalovic et al. 2011) (Table 1).

Clinical Implication of a Positive Gram Stain

There are five major clinical scenarios, i.e., pharyngitis, CAP, wound infections, ABM, and UTIs, when the Gram stain provides rapid and accurate information on differentiating non-infection, i.e., colonization (no antibiotic therapy) from infection (appropriate for initial empiric therapy). All of the significant clinical laboratory tests require interpretation based on clinical context. The Gram stain's strength in ASP is based not only on morphologically based DDx, but also on the associated white blood cell (PMNs vs. mononuclear cells) response. Otherwise unexplained, abundant bacteria plus numerous PMNs indicates infection, not colonization. In contrast, few to many bacteria with few/no PMNs indicates colonization not infection. Basically, the properly collected and interpreted Gram stain interpretation should take into account both the WBC response (Gram stain background) as well as the morphology/arrangement of the Gram stain bacterial pathogens. Traditionally, with ABM, the Gram stain has been used not only to differentiate bacterial from viral acute meningitis but also has been the basis for initial empiric antibiotic therapy based on CSF Gram stain morphology (Cunha and Cunha 2017, Bottone 2004, de la Maza et al. 2004, Cohen et al. 2016, Koneman et al. 1997).

Gram stain in wound infections

With wound Gram stains the key objective is to demonstrate a predominant organism, i.e., one pathogen (wound infection/cellulitis) vs. multiple pathogens (in an abscess) as the cause of infection. If the wound has a serous/serosanguinous drainage then the DDx is limited to GAS/GBS cellulitis (organisms seen on Gram stain) vs. colonization of a seroma (few/no organisms and few PMNs) seen on Gram stain. Clinically, if the wound discharge is purulent, infection is certain and the Gram stain shows the predominant morphologic type/arrangement of the pathogens pending wound culture (Cunha and Cunha 2017, Cunha et al. 2014) (Table 2).

Table 2. Post Surgical Draining Wounds: Colonization vs. Infection.

Wound Drainage	Wound Gram Stain	Wound Culture		Diagnosis
Clear	Few or some WBCs	+	→	Colonization
Serous	Few or some WBCs	+	→	Colonization
Serosanguinous	Few or some WBCs	+	→	Colonization
Purulent	Abundant WBCs	+	→	Infection

Adapted from: Cunha, B.A., E.D. Abruzzo and P.E. Schoch. 2014. Post-laminectomy wound infections: Colonized seromas mimicking wound infections. J. Clin. Med. 3: 191–196.

Gram stain in community acquired pneumonia (CAP)

If the sputum specimen is reflective of the lower respiratory tract (few/no epithelial cells) with abundant PMNs and a predominant morphologic pneumonia pathogen, the results are rapid, accurate and sufficient for presumptive diagnosis upon which to base initial empiric antibiotic therapy. In normal hosts, the DDx of CAP excluding AECB, includes *S. pneumoniae* (GPC in pairs), *H. influenzae* (pleomorphic GNB), *M. catarrhalis* (GN diplococci), *K. pneumoniae* (plump GNB), and MSSA/MRSA (GPC in clusters). Once again, clinical correlation helps select initial antibiotic therapy based on Gram stain results and host factors, e.g., CAP due to MSSA/MRSA occurs only in patients with influenza simultaneously, but not in normal hosts or even those with diabetes. GNB (encapsulated/plump) indicates *Klebsiella pneumoniae* CAP, but this occurs mostly in patients with history of alcoholism, not normal hosts. *P. aeruginosa* (thin/unencapsulated GNB) causes nosocomial or hospital acquired pneumonia (NAP/HAP), but not CAP. Even leukopenic hosts, who often develop *p. aeruginosa* bacteremia do not get *P. aeruginosa* CAP. Remaining in the DDx are *S. pneumoniae, H. influenzae* and *M. catarrhalis*. From an ASP perspective, treating any one of these possibilities with one antibiotic, e.g., ceftriaxone, covers all Gram stain determined possibilities (Cunha and Cunha 2017, Cunha 2004) (Table 3).

Table 3. Clinical Interpretation of the Sputum Gram Stain: Community Acquired Pneumonia (CAP).

Gram Stain	**Likely CAP Pathogen**	**Comments**
Gram positive *diplococci*	*S. pneumoniae*	Lancet shaped encapsulated diplococci (not streptococci)
Gram positive cocci (grape like clusters)	*S. aureus* (MSSA/MRSA)	Clusters predominant. Only MSSA/MRSA CAP with influenza
Gram positive cocci (short chains or pairs)	Group A *streptococci* (GAS)	Virulence inversely proportional to length of streptococci. Clusters not present
Gram positive (beaded/ filamentous branching)	Nocardia	Coccobacillary forms
Gram negative (coccobacillary)	*H. influenzae*	Pleomorphic and may be encapsulated
Gram negative bacilli	*Klebsiella pneumoniae* *P. aeruginosa*	Plump and encapsulated Thin and often arranged in end-to-end pairs
Gram negative *diplococci*	Moraxella (Branhamella) catarrhalis Neisseria meningitidis	Kidney bean shaped diplococci

Adapted from: Cunha, C.B. and B.A. Cunha. 2017. Antibiotic Essentials (15th ed.). JayPee Medical Publishers, New Delhi.

Gram stain in acute bacterial meningitis (ABM)

Once again, morphology/arrangement determines the microbiologic DDx. GPC in pairs indicates *S. pneumoniae*, whereas GPC in chains points to GBS. Gram negative diplococci indicate *N. meningitis*, while GNB are most likely to be *H. influenzae*. Gram positive bacilli indicate Listeria. Skin contaminants, e.g., diptheroids may be cultured from the CSF, but are not present in sufficient number to be seen on CSF Gram stain. Excluding, CSF shunt/drain related infection, MSSA, MRSA, CoNS and non-*H. influenzae* GNB are not usual ABM pathogens (Cunha and Cunha 2017, Bottone 2004, Cohen et al. 2016, Cunha 2013) (Table 4).

Table 4. Clinical Interpretation of the Gram Stain: CSF Acute Bacterial Meningitis (ABM).

Gram Stain	Likely CSF Pathogen	Comments
Gram positive bacilli	Listeria	Pseudomeningitis if due to Bacillus, Corynebacteria
Gram negative bacilli	*H. influenzae* (small, encapsulated, pleomorphic)	Enteric aerobic bacilli (larger, unencapsulated)
Gram positive cocci	Gp, A, B, D streptococci (pairs/chains) *S. pneumoniae* (pairs)	*S. aureus* (clusters) *S. epidermidis* (CoNS) (pairs/clusters)
Gram negative *diplococci*	Neisseria meningitidis	Gram stain may be negative
Mixed organisms	Aerobic/anaerobic organisms (brain abscess with meningeal leak) or *S. stercoralis* hyperinfecturm syndrome	Meningitis 2° (open head trauma)

Adapted from: Cunha, C.B. and B.A. Cunha. 2017. Antibiotic Essentials (15th ed.). JayPee Medical Publishers, New Delhi.

Gram stain in urinary tract infections (UTIs)

One of the most needed objectives in ASP is to dissuade physicians from treating asymptomatic bacteria (ASB). ASB, clinically represents, colonization of the urinary tract, and not infection. All too often, physicians, order urine cultures (UC) without obtaining microscopic examination as part of the urinalysis (UA). The degree of pyuria (microscopic, not by dipstick) determines the clinical significance of the bacteriuria on UC. If microscopic UA is ordered, the cellular "background" of the urine Gram stain is a guide to the clinical significance of UC results. It is the degree of pyuria, not bacterial colony count in UC, that determines clinical significance, i.e., ASB or UTI. Therefore, the Gram's stain, clinical utility in ASP is greater than ever in the current climate of rampant antibiotic empiricism. The Gram stain has a double benefit, i.e., not only does stating bacterial pathogen provide DDx possibility (in the proper clinical contest), but can help determine colonization vs. infection by also containing WBC (if one looks!) (Cunha and Cunha 2017, Hudepohl et al. 2016, Cunha 1981) (Tables 5, 6).

Gram stain in acute pharyngitis

With acute bacterial pharyngitis due to GAS, the Gram stain provides critical information on the local cellular response which microbiologically differentiates GAS colonization from infection in patients with acute pharyngitis. Rapid GAS testing or throat culture, does not differentiate GAS colonization from GAS infection. They only indicate that GAS is present

Table 5. Clinical Interpretation of the Urine Gram Stain: Acute Uncomplicated Cystitis (AUC) & Acute Pyelonephritis.

Gram Stain	Likely Uropathogen	Comments
Gram-positive cocci (clusters)	*S. aureus* *S. epidermidis* (CoNS) *S. saprophyticus*	Skin flora contaminant (not uropathogens) Uropathogen
Gram positive cocci (short chains)	Group B *streptococci* Group D *streptococci* *E. faecalis* (VSE) *E. faecium* (VRE)	Uropathogen* Uropathogen*
Gram negative bacilli	Coliform bacilli	Uropathogen*
Gram negative diplococci*	*N. gonorrhoeae* *N. meningitidis*	Gonococcal urethritis Rare cause of urethritis

* Depending upon the level of pyuria, may represent colonization or infection.
Adapted from: Cunha, C.B. and B.A. Cunha. 2017. Antibiotic Essentials (15th ed.). JayPee Medical Publishers, New Delhi.

Table 6. Acute Uncomplicated Cystitis (AUC) and Catheter Associated Bacteriuria (CAB)*: Colonization vs. Infection.

Inflammation

- Pyuria (> 30 WBCs/hpf) with no or low grade bacteriuria (< 50 cfu/ml).

Colonization

- AUC: bacteriuria with minimal pyuria (< 10 WBCs/hpf) and low grade bacteriuria (< 50 cfu/ml).
- CAB: bacteriuria with pyruia (> 10 WBCs/hpf) and bacteriuria (> 50 cfu/ml).

Infection

- AUC with moderate—high grade pyuria (> 10 WBCs/hpf) and high grade bacteriuria (> 100 cfu/ml).*
- CAB (UA/UC after Foley removed/replaced with > 10 WBCs/hpf and high grade bacteriuria (> 100 cfu/ml).*

* if indwelling urinary catheter, change/replace catheter and repeat AU/UC before considering treatment
Adapted from: Cunha, C.B. and B.A. Cunha. 2017. Antibiotic Essentials (15th ed.). JayPee Medical Publishers, New Delhi.

in the pharynx, not its clinical significance. It is well known, that ~ 30% of patients with acute EBV infectious mononucleosis pharyngitis are colonized with GAS. Rapid strep tests or GAS throat cultures do not indicate the significance of GAS, only the Gram stain can! The Gram stain of throat is not helpful in the diagnosis GAS pharyngitis (it cannot), but the Gram stain of the throat in GAS positive pharyngitis differentiates colonization (the physician should look for an underlying viral cause of pharyngitis) from infection by demonstrating the host's local WBC response in the pharynx. In *bona fide* GAS pharyngitis, the Gram stain shows abundant PMNs, and much cellular debris. In contrast, in GAS colonization of the pharynx, the Gram stain shows few/no PMNs, perhaps a few mononuclear cells and no cellular debris clearly indicates the role of GAS colonization in viral pharyngitis (Cunha and Cunha 2017, Bottone 2004, de la Maza et al. 2004, Cohen et al. 2016, Koneman et al. 1997).

Clinical Implications of a Negative Gram Stain

Gram stain results, like all other laboratory tests, must be interpreted in the appropriate clinical context. Combining Gram stain results with selective other key tests can further enhance the usefulness of Gram stain results. The diagnostic importance of negative Gram stain findings with proper specimen collection, blood sampling of the infected site/fluid and correct staining technique aside, may be as important as positive findings. Clinical examples include ABM, CAP, acute bacterial pharyngitis (ABP), wound infections and UTIs (Cunha and Cunha 2017, Bottone 2004, de la Maza et al. 2004, Cohen et al. 2016, Koneman et al. 1997, Madigan et al. 2003, Versalovic et al. 2011).

Acute bacterial meningitis (ABM)

With possible ABM, a negative Gram stain with a cloudy CSF suggests *S. pneumoniae* or *N. meningitis* as most likely pathogens. Autolysis of these pathogens may result in a negative Gram stain, but culture will later be positive. Also, even with a negative Gram stain in ABM due to the pneumococci or meningococci, CSF profile is pyogenic, i.e., elevated protein and a decreased CSF glucose. Furthermore, CSF lactic acid (LA) levels will be highly elevated (> nm/L). In cases of possible ABM with clear CSF and a negative Gram stain, a non-bacterial etiology is likely, e.g., viral, fungal, TB, drug induced. Importantly, the CSF LA levels are not elevated in viral and drug induced meningitis (Cunha and Cunha 2017, Cunha 2013, Cunha et al. 2007).

Community acquired pneumonia (CAP)

If the sputum specimen is representative of the lower respiratory tract (many PMNs, not epithelial cells), a negative Gram stain in CAP suggests a viral or atypical bacterial etiology. The Gram stain in Legionnaire's disease is negative, i.e., few/no PMNs with many mononuclear cells and no predominant bacterial type on Gram stain. Similar cellular responses on negative sputum on Gram stains are present with the other common atypical CAP, e.g., *M. pneumoniae*, *C. pneumoniae*, psitheosis, tularemia (Cunha and Cunha 2017, Cunha 2004, Cunha 2008).

Acute bacterial pharyngitis (ABP)

In ABP, a negative Gram stain, i.e., negative for evidence of bacterial pharyngitis (no abundance of PMNs, cellular debris, etc.). As previously noted, the clinical value of the Gram stain in pharyngitis is to differentiate GAS colonization from GAS pharyngitis is not based on the bacterial component of the Gram stain. Even if GAS demonstrated by rapid strep test or throat culture without a cellular background of infection (abundant PMNs with much cellular debris) such results represent GAS colonization, not infection (Cunha and Cunha 2017, Bottone 2004, de la Maza et al. 2004, Koneman et al. 1997, Jui et al. 1985, Sharma and Subbukrishnan 1981).

Urinary tract infections (UTIs)

In acute pyelonephritis (AP) and acute uncomplicated cystitis (AUC) the Gram stain, if positive, narrows the DDx of possible uropathogens to either GPC in chains (VSE, VRE, GBS) or aerobic GNB. A negative urine Gram stain effectively rules out AP and AUC. With AUC, the degree of pyuria on Gram stain differentiates colonization from infection. Regardless of the UC colony count, if there are < 10 WBC/hpf, the urine is colonized not infected. This important distinction cannot be too strongly stressed as a key component of ASP. Unnecessary antibiotic treatment of ASB results wasted resources possible side effects, and sets the stage for potential resistance (if "high resistance potential" antibiotics are used, e.g., TMP-SMX, ampicillin, ciprofloxacin) (Cunha and Cunha 2017, Hudepohl et al. 2016, Cunha 1981, Cunha 2001).

Summary

The Gram stain results, i.e., reflecting both the WBC component as well as the bacterial component provides important clinical information rapidly, accurately and inexpensively. More importantly, both positive and negative Gram stain results, interpreted in the proper clinical context, provides information not available even with molecular based diagnostic modalities. PCR and other molecular diagnostic methods accurately and expensively identify organisms, but do not help in determining their clinical significance for the patient. Clinical significance depends on clinical correlation based on clinical experience. In common and important clinical scenarios, the Gram stain's importance remains unsurpassed. Only the Gram stain provides critical information on both the WBC cellular response and limits DDx to the likely organism based on morphologic/arrangement. Properly applied and interpreted, the Gram stain is a key asset to ASP in reducing unnecessary antibiotic therapy of bacterial colonization as well as the needless treatment of non-bacterial, i.e., viral infections (Table 7).

Table 7. Antibiotic Stewardship Program (ASP) Gram Stain Directed Initial Empiric Therapy.

Clinical Scenarios	**Gram Stain Findings**		**Interpretation of Gram Stain Results**	**ASP Empiric Initial Antibiotic Therapy PCN tolerant Penicillin Allergic (PA)**
	Many PMNs	**Morphology/ Arrangement**	**Microbiologic Diagnosis**	
Acute Pharyngitis	+	GPC (in chains)	GAS pharyngitis	Amoxicillin PA: clindamycin
	–	GPC (in chains)	Viral pharyngitis	No antibiotic therapy Dx tests for viral pharyngitis
Community Acquired Pneumonia (CAP)	+	GPC (in pairs)	*S. pneumoniae*	Ceftriaxone PA: Respiratory quinolone or doxycycline
	+	GNB (pleomorphic)	*H. influenzae*	Ceftriaxone PA: Respiratory quinolone or doxycycline
	+	GND (diplococci)	*B. catarrhalis*	Ceftriaxone PA: Respiratory quinolone or doxycycline
	–	No organisms seen	Atypical CAP	Doxycycline or Levofloxacin alternatively Azithromycin
Acute Bacterial Meningitis (ABM)	+	GPB	*Listeria monocytogenes*	PA: Ampicillin (meningeal dosed) TMP-SMX or chloramphenicol
	+	GPC (pairs)	*S. pneumoniae*	Ceftriaxone PA: Meropenem (meningeal dosed)
	+	GND (diplococci)	*N. meningitis*	Ceftriaxone PA: Meropenem (meningeal dosed)
	+	GNB (pleomorphic)	*H. influenzae*	Ceftriaxone PA: Meropenem (meningeal dosed)
	–	No organisms seen	Aseptic meningitis	CSF tests for CNS viral pathogens (HSV, enteroviruses, etc.) If HSV ½ and VZV negative, no antibiotic or antiviral therapy

Wounds	+	GPC (in chains)	GAS/GBS cellulitis	Cefazolin PA: Clindamycin
	+	GPC (in clusters)	MSSA/MRSA abscess	Vancomycin Alternatively: Minocycline
	–	No organisms seen	Colonized seroma	No antibiotic therapy needed
Urinary Tract Infections (UTIs)	+	GPC (in chains)	VSE/VRE	Nitrofurantoin Alternately: Amoxicillin
	+	GNB	Coliforms	Amoxicillin Alternately: Cephalexin PA: Nitrofurantoin
	–	GPC or GNB	UT colonization	No antibiotic therapy
	+	Few organisms seen	UT inflammation	No antibiotic therapy
	+	GPC (in chains)	VSE	Piperacillin/tazobactam PA: Meropenem
			VRE	Linezolid Alternately: Daptomycin
	+	GNB	Coliforms	Ceftriaxone PA: Azithromycin
	–	Few organisms seen	AP unlikely	No antibiotic therapy

AP: acute pyelo nephritis

GND: Gram negative diplococci

GNB: Gram negative bacilli

GPC: Gram positive cocci

GAS: Group A Streptococci

GBS: Group B Streptococci

MSSA: Methicillin sensitive *Staphylococcus aureus*

MRSA: Methicillin resistant *Staphylococcus aureus*

References

Bottone, E.J. (ed.). 2004. An Atlas of the Clinical Microbiology of Infectious Disease. Volume 1. Bacterial Agents. The Parthenon Publishing Group, Boca Raton.

Cohen, J., W.G. Powderly and S. Opal (eds.). 2016. Infectious Diseases (4th ed.). Elsevier, Philadelphia.

Cunha, B.A. 1981. Urinary tract infections. 1. Pathophysiology and diagnostic approach. Postgrad. Med. 70: 141–150.

Cunha, B.A. 2000. Antibiotic resistance. Med. Clin. North America 84: 1407–1429.

Cunha, B.A. 2001. Effective antibiotic resistance control strategies. Lancet 357: 1307–1308.

Cunha, B.A. 2004. Emperic therapy of community acquired pneumonia: guidelines for the perplexed? Chest. 125: 1913–1919.

Cunha, B.A., R. Fatehpuria and L.E. Eisenstein. 2007. Listeria monocytogenes encephalitis mimicking herpes simplex virus encephalitis: The differential diagnostic importance of cerebrospinal fluid lactic acid levels. Heart & Lung 36: 226–231.

Cunha, B.A. 2008. Atypical pneumonias: Current clinical concepts focusing on Legionnaires' disease. Curr. Opin. Pulm. Med. 14: 183–194.

Cunha, C.B., C.A. Varughese and E. Mylonakis. 2013. Antimicrobial stewardship programs (ASPs): The devil is in the details? Virulence 15: 147–149.

Cunha, B.A. 2013. The clinical and laboratory diagnosis of acute meningitis and acute encephalitis. Expert Opin. Med. Diagn. 7: 343–364.

Cunha, B.A., E.D. Abruzzo and P.E. Schoch. 2014. Post-laminectomy wound infections: Colonized seromas mimicking wound infections. J. Clin. Med. 3: 191–196.

Cunha, C.B. and B.A. Cunha (eds.). 2017. Antibiotic Essentials (15th ed.). JayPee Medical Publishers, New Delhi, pp. 1–16, 185–253.

de la Maza, L.M., M.T. Pezzlo, J.T. Shigei, G.L. Tan and E.M. Peterson (eds.). 2004. Color Atlas of Medical Bacteriology. ASM Press, Washington, DC.

Hudepohl, N.J., C.B. Cunha and L.A. Mermel. 2016. Antibiotic prescribing for urinary tract infections in the emergency department based on local antibiotic resistance patterns: implications for antimicrobial stewardship. Infect. Control Hosp. Epidemiol. 37: 359–360.

Jui, J., R. Norton, S. Edminster and S. Boyer. 1985. Gram stain for streptococcal pharyngitis. Ann. Emerg. Med. 14: 191–192.

Koneman, E.W., S.W. Allen, W.M. Janda, P.C. Schreckenberger and W.C. Winn (eds.). 1997. Color Atlas and Textbook of Diagnostic of Diagnostic Microbiology (5th ed.). Lippincott-Raven Publishers, Philadelphia.

Madigan, M.T., J.M. Martinko and J. Parker (eds.). 2003. Brock Biology of Microorganisms (10th ed.), Prentice Hall, Upper Saddle River, NJ.

Munoz-Gomez, S., E. Wirkowski and B.A. Cunha. 2015. Post craniotomy extra-ventricular drain (EVD) associated nosocomial meningitis: CSF diagnostic criteria. Heart Lung 44: 158–160.

Sharma, S.C. and P.V. Subbukrishnan. 1981. Streptococcal pharyngitis—rapid diagnosis by Gram stain. Postgrad. Med. J. 57: 13–15.

Versalovic, J., K.C. Carroll, G. Funke, J.M. Jorgensen, M.L. Landry and D.W. Warnock (eds.). 2011.Manual of Clinical Microbiology (10th ed.). ASM Press, Washington, DC.

CHAPTER 5

Microbiological Diagnosis of Tuberculosis

Hitender Gautam[1,*] and *Urvashi B. Singh*[2]

INTRODUCTION

Tuberculosis disease is one of the foremost causes of morbidity and mortality in world. As reported to World Health Organization (WHO), the estimated number of new cases of tuberculosis was 8.6 million in 2014 and the number of deaths approximately 1.5 million (WHO 2015). If one disease has maintained the status of global emergency even after worldwide efforts to control it, its tuberculosis.

During the past two decades, the increase in drug-resistant tuberculosis has added to the complexity of this disease especially multidrug resistant tuberculosis (MDR-TB) and extensively drug resistant tuberculosis (XDR-TB) which are much more challenging to treat than drug-susceptible disease (Gandhi et al. 2006). The estimated number of MDR-TB cases worldwide is approximately 5,00,000 described from at least 127 countries, and XDR-TB has been described from 105 countries (WHO 2015).

Mycobacterium tuberculosis spreads from person to person through the airborne route. Numerous aspects regulate the likelihood of *Mycobacterium tuberculosis* transmission: (1) level of infectivity of source patient has a strong association with transmission especially acid-fast bacilli (AFB) seen on sputum smear and/or chest radiograph with cavitary lesions; (2) susceptibility (immune status) of the contact; (3) period of exposure

[1] Assistant Professor, Department of Microbiology, All India Institute of Medical Sciences (AIIMS), New Delhi, India.

[2] Professor, Department of Microbiology, All India Institute of Medical Sciences (AIIMS), New Delhi, India.

* Corresponding author: drhitender@gmail.com

of the source patient to contact; (4) the settings in which exposure takes place (a small, inadequately ventilated area carries the maximum risk); and (5) virulence of *Mycobacterium tuberculosis* strain (Lewinsohn et al. 2016).

Extrapulmonary tuberculosis (EPTB) comprises approximately 15 to 20 percent of tuberculosis cases worldwide (WHO, SEAR 2015). Reaching a diagnosis in EPTB cases is always challenging. Due to relatively inaccessible sites for clinical sample collection and pauci-bacillary nature of samples collected, sensitivity of the diagnostic tests is low. To begin with, cell counts and chemistries should be done on fluid samples obtained from suspected sites in cases of extrapulmonary tuberculosis. Samples on which cell counts and chemistries are acceptable include pleural fluid, ascitic fluid, cerebrospinal fluid, and various joint fluids. Cell counts and chemistries have the advantage that they can be performed in hours, are economical, and are technically easy. Sensitivity is moderate to high, but there are issues with the specificity especially if interpreted alone. The specificity of the counts and chemistries is much improved when interpreted in association with clinical findings, radiographic picture, and further laboratory investigations.

Although a number of tests are available for diagnosis of pulmonary and extra-pulmonary tuberculosis, efficient tuberculosis (TB) control is hampered by a real lack of a rapid field screening assays (Boehme et al. 2005). In addition, currently available tests have varied sensitivity, specificity and clinical implications. Successful tuberculosis control requires a system of public and private sector coordination of laboratories to standardise investigations and the flow of information. These laboratory investigations should be accessible to clinicians involved in tuberculosis diagnosis and management promptly.

Respiratory Specimens in Suspected Pulmonary Tuberculosis

Sputum collected by spontaneous expectoration is the preferred specimen in adults. Respiratory specimens that can be collected from children include gastric aspirates, sputum collected by spontaneous expectoration or induction, broncho-alveolar lavage (BAL) and nasopharyngeal aspiration. Gastric aspirate needs to be collected on three consecutive mornings in patients with suspected pulmonary tuberculosis. It offers an overall diagnostic yield of up to 40–50 percent, with higher positivity for infants (approximately 90 percent), symptomatic children, and children who have extensive disease (up to 75 percent), when used for confirmation of clinical diagnosis of tuberculosis disease in low prevalence populations (Loeffler 2003, Marais et al. 2006). Nasopharyngeal aspiration or induced sputum from children has shown a yield of 20–30 percent (Hatherill et al. 2009), whereas broncho-alveolar lavage (BAL) has shown a yield of 10–20 percent in pediatric population with suspected pulmonary tuberculosis (Abadco and Steiner 1992). In case of negative results, clinicians often treat TB presumptively in pediatric patients in the appropriate clinical setting.

Alternate samples are needed in adults suspected of pulmonary tuberculosis and negative sputum smear microscopy and for those unable to produce sputum. Studies have reported a higher yield from induced sputum than bronchoscopy in such cases (Conde et al. 2000, Saglam et al. 2005, Brown et al. 2007). The diagnostic yield of induced sputum rises with multiple specimens, with detection rates of more than 90 percent for AFB smear microscopy and more than 98 percent for mycobacterial culture when three or more specimens are processed (Charoenratanakul et al. 1995). Sputum induction is less costly in comparison to bronchoscopy (Anderson et al. 1995, McWilliams et al. 2002). It has equal or greater diagnostic value than bronchoscopic sampling, has less risk, and is preferred as a primary respiratory sampling technique. Nevertheless, there is a potential advantage to bronchoscopy with biopsy over sputum induction in getting to an early presumptive diagnosis of TB based on histopathologic findings typical of tuberculosis. Overall, bronchoscopic sampling has a diagnostic yield ranging from 50 to up to 100 percent when it is based on histopathology and culture in suspected pulmonary tuberculosis cases.

Transbronchial biopsy (TBB) has shown histopathologic findings indicative of pulmonary tuberculosis in 40–60 percent of specimens from smear-negative HIV-uninfected patients (Miro et al. 1992, Kennedy et al. 1992).

Bronchoscopic sampling has clear advantage in cases with suspected pulmonary tuberculosis when respiratory samples can not be obtained by non-invasive procedures or are non-revealing.

Although not routinely done, post-bronchoscopy sputum acid fast bacilli (AFB) smears have a diagnostic positivity of 10–70 percent and post-bronchoscopy mycobacterial cultures have a positivity of 35–70 percent (Wallace et al. 1981, Sarkar et al. 1980, Schoch et al. 2007).

Miliary tuberculosis needs to be differentiated from a number of other pulmonary diseases. Bronchoscopic sampling including a biopsy should be done early in the work up.

Acid fast bacilli (AFB) smear microscopy in the diagnosis of pulmonary tuberculosis

AFB smear still forms the back bone of tuberculosis diagnosis. Carrying out three acid fast (AFB) smears diagnose pulmonary tuberculosis with a sensitivity of nearly 70 percent when culture-confirmed tuberculosis disease is used as the reference standard (Lewinsohn et al. 2016). Each specimen escalates sensitivity which forms the basis for carrying out three AFB smears routinely. AFB smear with single specimen has sensitivity of approximately 50 percent.

The sensitivity of a first morning specimen is 12 percent better than a single random specimen (Mase et al. 2007). When we use culture as the standard, microscopy on concentrated specimens has a mean rise in sensitivity of 18 percent compared with non-concentrated specimens and fluorescence microscopy is around 10 percent further sensitive in comparison to conventional microscopy (Steingart et al. 2006a, 2006b). The specificity of microscopy is reasonably high (≥ 90 percent), but the positive predictive value (PPV) varies (70–90 percent) depending on the prevalence of tuberculosis versus non-tuberculous mycobacterial disease (Gordin and Slutkin 1990, Yajko et al. 1994).

The obvious advantages of AFB smear microscopy are its technical simplicity and quick results. To improve the yield, 5–10 ml of sputum sample is ideal and at least 3 ml should be collected.

AFB smear microscopy in the diagnosis of extra-pulmonary tuberculosis

Due to the paucibacillary nature of extrapulmonary tuberculosis specimens, the diagnostic capability and sensitivity of smear microscopy for acid fast bacilli (AFB) is less reliable in extrapulmonary tuberculosis than pulmonary tuberculosis. As per the literature, AFB smear microscopy has a sensitivity ranging from 0–10 percent, 14–39 percent, 10–30 percent, < 5 percent, and 0–42 percent in pleural fluid, pleural tissue, cerebrospinal fluid, peritoneal and pericardial fluid, respectively (Epstein et al. 1987, Chan et al. 1991, Berenguer et al. 1992, Ogawa et al. 1987, Haas et al. 1977, Monteyne and Sindic 1995, Karney et al. 1977, Shakil et al. 1996, Sherman et al. 1980, Singh et al. 1969, Quale et al. 1987, Fowler and Manitsas 1973), considering the combination of clinical, pathological, cytological and microbiologic diagnosis as the confirmatory standard. But, specificity of acid fast bacilli (AFB) smear microscopy is ≥ 90 percent as described for pulmonary tuberculosis. In view of the above, a positive result has a significant role in diagnosis of extrapulmonary tuberculosis and can guide decision making especially when false-positive results are not likely. On the other hand, a negative result is difficult to be used to rule out tuberculosis in the context of low sensitivity.

Mycobacterial cultures in the diagnosis of pulmonary tuberculosis

Reference standard for tuberculosis diagnosis in the laboratory is mycobacterial culture. Cultures in liquid media have reasonable sensitivity and high specificity, but potential contamination is a limitation. Cultures on solid media alone are not adequately sensitive to reliably diagnose tuberculosis and in general take longer to produce results. Use of both liquid and solid media increases the sensitivity of mycobacterial cultures, with solid media cultures offering defense against contamination and the liquid media providing faster results.

A meta-analysis showed, liquid mycobacterial culture methods to be more sensitive (88 to 90 percent) than the solid mycobacterial culture methods (75 percent) with a shorter time to detection (13.2 to 15.2 days for liquid mycobacterial culture methods versus 25.8 days for the solid mycobacterial culture method). The specificity of both liquid and solid mycobacterial culture methods surpasses 99 percent. Liquid culture medium has a higher contamination rate than solid culture medium due to the growth of bacteria other than mycobacteria (4–9 percent) which hampers with obtaining a valid culture result (Crucian et al. 2004).

For optimal results, both liquid and solid media mycobacterial cultures are used routinely.

Mycobacterial cultures in the diagnosis of extrapulmonary tuberculosis

Published studies have shown that mycobacterial culture has a sensitivity ranging from 23–58 percent, 40–58 percent, 45–70 percent, 80–90 percent, 45–69 percent, and 50–65 percent in pleural fluid, pleural tissue, cerebrospinal fluid, urine, peritoneal fluid, and pericardial fluid, respectively (Epstein et al. 1987, Chan et al. 1991, Haas et al. 1977, Karney et al. 1977, Shakil et al. 1996, Sherman et al. 1980, Singh et al. 1969, Berger and Mejia 1973, Seibert et al. 1991, Simon et al. 1977, Christensen 1974, Alvarez and McCabe 1984), considering the combination of clinical, pathological, cytological and microbiologic diagnosis as the standard. The rate of false positivity is less than 3 percent for extrapulmonary mycobacterial culture which signifies the importance of positive result. Since false-negative results are quite frequent, additional diagnostic testing is always indicated.

Nucleic acid amplification tests (NAAT) for pulmonary tuberculosis

Mycobacterial cultures need at least 1–2 week and often longer to start showing growth. For early and speedy diagnosis, NAAT has benefits over smear microscopy by distinguishing *Mycobacterium tuberculosis* and non-tuberculous mycobacteria (NTM). NAAT should be used in addition to mycobacterial culture and smear microscopy as NAAT is not sensitive enough to substitute for mycobacterial culture for diagnosis and it also does not produce an isolate, which is an absolute requirement for phenotypic drug susceptibility testing (DST). The second issue is that the test characteristics of NAAT are inconsistent depending upon the AFB smear result and clinical suspicion. Among smear-positive specimens, NAAT yields false-negative results in only approximately 5% cases making it reliable for excluding pulmonary tuberculosis in this setting. When there is a high clinical index of suspicion for disease, the specificity of NAAT is high and a positive result can be used as presumptive evidence of tuberculosis and guide therapeutic decisions. However, due to low sensitivity, it's difficult to exclude pulmonary tuberculosis disease on the basis of NAAT only. When the clinical suspicion for tuberculosis is low, false-positive NAAT results are unacceptably frequent and therefore not reliable (Shingadia and Novelli 2003). NAAT has added to the armamentarium of tests currently available for diagnosis of tuberculosis especially in the presence of high likelihood of clinical diagnosis. Studies have also found an increase in the usefulness of NAAT tests if applied on multiple samples rather than a single specimen (Moore et al. 2005, Campos et al. 2008). Presence of inhibitors can reduce the sensitivity for NAATs (Sloutsky et al. 2004) in single specimens. Recommendation for the use of NAATs is based on commercial test kits. For in-house tests, greater heterogeneity remains a drawback (Flores et al. 2005). If in-house tests are to be used in the laboratory, they require to be validated and should have analytical performance accuracy comparable to commercial tests. NAAT tests have shown corroborative value with AFB smear microscopy. With positive AFB smear microscopy, NAAT has a sensitivity of approximately 96 percent and specificity of 85 percent, respectively. In case of negative AFB smear microscopy, the sensitivity decreases to 66 percent and the specificity improves to 98 percent. Prior antimycobacterial therapy has a significant impact on the test result.

NAAT for extrapulmonary tuberculosis

Nucleic acid amplification used on pleural fluid and cerebrospinal fluid has a sensitivity of approximately 56 percent and 62 percent, respectively. In comparison, specificity of NAAT for both pleural and cerebrospinal fluid approximately 98 percent for both (Pai et al. 2004, Pai et al. 2003). These results favor the use of NAAT on suspected extrapulmonary tuberculosis samples.

Rapid molecular drug susceptibility testing (DST)

Rapid molecular drug susceptibility testing is useful for early and individualized anti-tuberculosis therapy. Molecular beacon assays and line probe assays have been used for rapid molecular DST. The sensitivity has been found to be ≥ 97 percent and specificity ≥ 98 percent for detection of rifampin resistance with line probe assays when compared to standard of conventional culture-based DST. Xpert MTB/RIF, a molecular-beacon dependent technique for quick rifampin resistance detection has shown sensitivity of > 92 percent and specificity of > 99 percent for recognition of rifampin resistance in a single specimen. The sensitivity improves to > 97 percent when used on three specimens (Boehme et al. 2010). In populations with a low prevalence of drug resistance, the positive predictive value (PPV) for the recognition of rifampin resistance is low (CDC 2013). For isoniazid resistance detection in rapid molecular DST, sensitivity and specificity is of the order of 90 percent and 99 percent, respectively, in comparison to culture-based DST as the reference standard.

Conventional culture-based DST is still the laboratory reference standard for tuberculosis and should be performed routinely any time *Mycobacterium tuberculosis* complex is detected in culture (Woods et al. 2011). It is even more critical for successful treatment of patients with MDR-TB. Treatment success can reach 75% or more (Chan et al. 2004, Narita et al. 2001) and is dependent directly on patients being treated with an effective anti-tuberculosis regimen (Lew et al. 2008). The significant constraint of culture-based DST is that it can take > 2 weeks to grow the isolate that is essential for testing and can delay the results. Rapid molecular DST can be completed within hours, facilitating earlier commencement of an effective antimicrobial regimen.

As, the sensitivity and specificity of rapid molecular DST for identifying rifampin resistance are both > 97 percent, it can be relied upon to confirm or rule-out resistance to rifampin among respiratory specimens. This scenario is not so clear for identifying isoniazid resistance, as true positivity and true negativity occurs in approximately 90 percent and 99 percent respectively for rapid molecular DST. This leaves around 1% and 10% of the false-positive and false-negative results, respectively which signifies the role of rapid molecular DST in confirming isoniazid resistance but not for its exclusion.

Adenosine deaminase (ADA)

Adenosine deaminase enzyme changes adenosine to inosine, and also deoxyadenosine to deoxyinosine in purine catabolism pathway, and thus catalyses irreversible deamination. ADA also has a role in proliferation along with differentiation of T-lymphocytes (Baganha et al. 1990). For pleural fluid, meta-analysis studies estimated that the sensitivity and specificity of an increased Adenosine deaminase (ADA) level in the pleural fluid are 89–99 percent and 88–97 percent, respectively. Thresholds which have been used to characterize an elevated ADA level ranged from 10 U/L to 71 U/L, with the majority clustering around 40 U/L (Morisson and Neves 2008, Liang et al. 2008, Greco et al. 2003, Goto et al. 2003). The pleural fluid ADA levels in pleural effusion due to tuberculosis are considerably higher in comparison to transudates (Chander and Shrestha 2012). For pericardial fluid, meta-analysis studies have estimated that the sensitivity and specificity of an increased Adenosine deaminase (ADA) level in pericardial fluid are 88 percent and 83 percent, respectively (Tuon et al. 2006). The threshold which has been used to characterize an elevated ADA level is 40 U/L. For peritoneal fluid, meta-analysis studies estimated that the sensitivity and specificity of an increased Adenosine deaminase (ADA) levels are 100 percent and 97 percent, respectively (Riquelme et al. 2006). Thresholds which were used to characterize an elevated ADA level ranged from 36 U/L to 40 U/L. Adenosine deaminase (ADA) levels should be measured, on fluid samples collected from patients with suspected pleural tuberculosis, tubercular meningitis, peritoneal tuberculosis, or pericardial tuberculosis.

Histopathology in the diagnosis of extrapulmonary tuberculosis

Histological examination has shown a sensitivity ranging from 70–95 percent, 85–94 percent, 80–100 percent, 60–70 percent and 75–100 percent in pleural tissue, urologic tissue, peritoneal biopsy, endometrial curettage and pericardial tissue, respectively, when using the combination of clinical, pathological, cytological and microbiologic diagnosis as confirmatory standard. There are concerns with specificity which tends to be lower because necrotizing and non-necrotizing granulomas are also observed in other infectious and non-infectious diseases. Another concern is that both positive and negative outcomes should be interpreted in the context of clinical scenario as neither false-positive nor false-negative results are uncommon.

Summary

Ideally, a laboratory test should be easy to perform, economical and rapid with high accuracy, i.e., > 95 percent sensitivity and specificity. Inspite of advances, the current rapid tests for tuberculosis diagnosis have less than satisfactory sensitivity and may be unsuccessful in identifying paucibacillary pulmonary tuberculosis. In addition, rapid tests are costly, which makes it challenging to apply them in areas with high tuberculosis prevalence and resource limited settings. Rapid tests for recognition of drug resistance are becoming more useful especially for rifampin. But issues of cost and requirement for drug resistance detection to other anti-tuberculosis agents remain.

Other areas that require attention and improvement are diagnosis of pediatric and extrapulmonary tuberculosis. Pediatric tuberculosis is more difficult to diagnose due to the lower yield of AFB smear and culture as is the case with extrapulmonary tuberculosis. Also there is a need for tests to follow and monitor response to treatment of tuberculosis. Management of HIV disease where both host and viral markers are available for monitoring response to drug therapy serves as an optimal example.

References

Abadco, D.L. and P. Steiner. 1992. Gastric lavage is better than bronchoalveolar lavage for isolation of *Mycobacterium tuberculosis* in childhood pulmonary tuberculosis. Pediatr. Infect. Dis. J. 11: 735–738.

Alvarez, S. and W.R. McCabe. 1984. Extrapulmonary tuberculosis revisited: A review of experience at Boston City and other hospitals. Medicine (Baltimore) 63: 25–55.

Anderson, C., N. Inhaber and D. Menzies. 1995. Comparison of sputum induction with fiber-optic bronchoscopy in the diagnosis of tuberculosis. Am. J. Respir. Crit. Care. Med. 152(5 pt 1): 1570–1574.

Baganha, M., A. Pego and M.A. Lima. 1990. Serum and pleural adenosine deaminase: Correlation with lymphocytic populations. Chest 97: 605–610.

Berenguer, J., S. Moreno, F. Laguna, T. Vicente, M. Adrados, A. Ortega et al. 1992. Tuberculous meningitis in patients infected with the human immunodeficiency virus. N. Engl. J. Med. 326: 668–672.

Berger, H.W. and E. Mejia. 1973. Tuberculous pleurisy. Chest 63: 88–92.

Boehme, C., E. Molokova, F. Minja, S. Geis, T. Loscher, L. Maboko et al. 2005. Detection of mycobacterial lipoarabinomannan with an antigen-capture ELISA in unprocessed urine of Tanzanian patients with suspected tuberculosis. Trans. R. Soc. Trop. Med. Hyg. 99: 893–900.

Boehme, C.C., P. Nabeta, D. Hillemann, M.P. Nicol, S. Shenai, F. Krapp et al. 2010. Rapid molecular detection of tuberculosis and rifampin resistance. N. Engl. J. Med. 363: 1005–1015.

Brown, M., H. Varia, P. Bassett, R.N. Davidson, R. Wall and G. Pasvol. 2007. Prospective study of sputum induction, gastric washing, and bronchoalveolar lavage for the diagnosis of pulmonary tuberculosis in patients who are unable to expectorate. Clin. Infect. Dis. 44: 1415–1420.

Campos, M., A. Quartin, E. Mendes, A. Abreu, S. Gurevich, L. Echarte et al. 2008. Feasibility of shortening respiratory isolation with a single sputum nucleic acid amplification test. Am. J. Respir. Crit. Care. Med. 178: 300–305.

CDC. 2013. Centers for Disease Control and Prevention. Availability of an assay for detecting *Mycobacterium tuberculosis*, including rifampin-resistant strains, and considerations for its use—United States. MMWR Morb. Mortal. Wkly. Rep. 62: 821–827.

Chan, C.H., M. Arnold, C.Y. Chan, T.W. Mak and G.B. Hoheisel. 1991. Clinical and pathological features of tuberculous pleural effusion and its long-term consequences. Respiration 58(3–4): 171–175.

Chan, E.D., V. Laurel, M.J. Strand, J.F. Chan, M.L. Huynh, M. Goble et al. 2004. Treatment and outcome analysis of 205 patients with multidrug-resistant tuberculosis. Am. J. Respir. Crit. Care. Med. 169: 1103–1109.

Chander, A. and C.D. Shrestha. 2012. Adenosine deaminase (ADA) analysis and its diagnostic role in tuberculous pleural effusion. Intl. J. Basic. and Applied Med. Sci. 2: 103–108.

Charoenratanakul, S., W. Dejsomritrutai and A. Chaiprasert. 1995. Diagnostic role of fiberoptic bronchoscopy in suspected smear negative pulmonary tuberculosis. Respir. Med. 89: 621–623.

Christensen, W.I. 1974. Genitourinary tuberculosis: review of 102 cases. Medicine (Baltimore) 53: 377–390.

Conde, M.B., S.L. Soares, F.C. Mello, V.M. Rezende, L.L. Almeida, A.L. Reingold et al. 2000. Comparison of sputum induction with fiberoptic bronchoscopy in the diagnosis of tuberculosis: experience at an acquired immune deficiency syndrome reference center in Rio de Janeiro, Brazil. Am. J. Respir. Crit. Care. Med. 162: 2238–2240.

Crucian, M., C. Scarparo, M. Malena, O. Bosco, G. Serpelloni and C. Mengoli. 2004. Meta-analysis of BACTEC MGIT 960 and BACTEC 460 TB, with or without solid media, for detection of mycobacteria. J. Clin. Microbiol. 42: 2321–2325.

Epstein, D.M., L.R. Kline, S.M. Albelda and W.T. Miller. 1987. Tuberculous pleural effusions. Chest. 91: 106–109.

Flores, L.L., M. Pai, J.M. Jr. Colford and L.W. Riley. 2005. In-house nucleic acid amplification tests for the detection of *Mycobacterium tuberculosis* in sputum specimens: meta-analysis and meta-regression. BMC Microbiol. 5: 55.

Fowler, N.O. and G.T. Manitsas. 1973. Infectious pericarditis. ProgCardiovasc. Dis. 16: 323–336.

Gandhi, N.R., A. Moll, A.W. Sturm, R. Pawinski, T. Govender, U. Lalloo et al. 2006. Extensively drug-resistant tuberculosis as a cause of death in patients co-infected with tuberculosis and HIV in a rural area of South Africa. Lancet. 368: 1575–1580.

Gordin, F. and G. Slutkin. 1990. The validity of acid-fast smears in the diagnosis of pulmonary tuberculosis. Arch. Pathol. Lab. Med. 114: 1025–1027.

Goto, M., Y. Noguchi, H. Koyama, K. Hira, T. Shimbo and T. Fukui. 2003. Diagnostic value of adenosine deaminase in tuberculous pleural effusion: a meta-analysis. Ann. Clin. Biochem. 40(pt 4): 374–381.

Greco, S., E. Girardi, R. Masciangelo, G.B. Capoccetta and C. Saltini. 2003. Adenosine deaminase and interferon gamma measurements for the diagnosis of tuberculous pleurisy: a meta-analysis. Int. J. Tuberc. Lung. Dis. 7: 777–786.

Haas, E.J., T. Madhavan, E.L. Quinn, F. Cox, E. Fisher and K. Burch. 1977. Tuberculous meningitis in an urban general hospital. Arch. Intern. Med. 137: 1518–1521.

Hatherill, M., T. Hawkridge, H.J. Zar, A. Whitelaw, M. Tameris, L. Workman et al. 2009. Induced sputum or gastric lavage for community-based diagnosis of childhood pulmonary tuberculosis? Arch. Dis. Child. 94: 195–201.

Karney, W.W., J.M. O'Donoghue, J.H. Ostrow, K.K. Holmes and H.N. Beaty. 1977. The spectrum of tuberculous peritonitis. Chest. 72: 310–315.

Kennedy, D.J., W.P. Lewis and P.F. Barnes. 1992. Yield of bronchoscopy for the diagnosis of tuberculosis in patients with human immunodeficiency virus infection. Chest. 102: 1040–1044.

Lew, W., M. Pai, O. Oxlade, D. Martin and D. Menzies. 2008. Initial drug resistance and tuberculosis treatment outcomes: systematic review and meta-analysis. Ann. Intern. Med. 149: 123–134.

Lewinsohn, D.M., M.K. Leonard, P.A. LoBue, D.L. Cohn, C.L. Daley, E. Desmond et al. 2017. Official american thoracic society/infectious diseases society of America/Centers for disease control and prevention clinical practice guidelines: Diagnosis of tuberculosis in adults and children. Clin. Infect. Dis. 64: e1–e33.

Liang, Q.L., H.Z. Shi, K. Wang, S.M. Qin and X.J. Qin. 2008. Diagnostic accuracy of adenosine deaminase in tuberculous pleurisy: a meta-analysis. Respir. Med. 102: 744–754.

Loeffler, A.M. 2003. Pediatric tuberculosis. Semin. Respir. Infect. 18: 272–291.

Marais, B.J., R.P. Gie, H.S. Schaaf, N. Beyers, P.R. Donald and J.R. Starke. 2006. Childhood pulmonary tuberculosis: Old wisdom and new challenges. Am. J. Respir. Crit. Care. Med. 173: 1078–1090.

Mase, S.R., A. Ramsay, V. Ng, M. Henry, P.C. Hopewell, J. Cunningham et al. 2007. Yield of serial sputum specimen examinations in the diagnosis of pulmonary tuberculosis: A systematic review. Int. J. Tuberc. Lung. Dis. 11: 485–495.

McWilliams, T., A.U. Wells, A.C. Harrison, S. Lindstrom, R.J. Cameron and E. Foskin. 2002. Induced sputum and bronchoscopy in the diagnosis of pulmonary tuberculosis. Thorax 57: 1010–1014.

Miro, A.M., E. Gibilara, S. Powell and S.L. Kamholz. 1992. The role of fiberoptic bronchoscopy for diagnosis of pulmonary tuberculosis in patients at risk for AIDS. Chest 101: 1211–1214.

Monteyne, P. and C.J. Sindic. 1995. The diagnosis of tuberculous meningitis. Acta. Neurol. Belg. 95: 80–87.

Moore, D.F., J.A. Guzman and L.T. Mikhail. 2005. Reduction in turnaround time for laboratory diagnosis of pulmonary tuberculosis by routine use of a nucleic acid amplification test. Diagn. Microbiol. Infect. Dis. 52: 247–254.

Morisson, P. and D.D. Neves. 2008. Evaluation of adenosine deaminase in the diagnosis of pleural tuberculosis: a Brazilian meta-analysis. J. Bras. Pneumol. 34: 217–224.

Narita, M., P. Alonso, M. Lauzardo, E.S. Hollender, A.E. Pitchenik and D. Ashkin. 2001. Treatment experience of multidrug-resistant tuberculosis in Florida, 1994–1997. Chest 120: 343–348.

Ogawa, S.K., M.A. Smith, D.J. Brennessel and F.D. Lowy. 1987. Tuberculous meningitis in an urban medical center. Medicine (Baltimore) 66: 317–326.

Pai, M., L.L. Flores, N. Pai, A. Hubbard, L.W. Riley and J.M. Colford Jr. 2003. Diagnostic accuracy of nucleic acid amplification tests for tuberculous meningitis: A systematic review and meta-analysis. Lancet Infect. Dis. 3: 633–643.

Pai, M., L.L. Flores, A. Hubbard, L.W. Riley and J.M. Colford Jr. 2004. Nucleic acid amplification tests in the diagnosis of tuberculous pleuritis: A systematic review and meta-analysis. BMC. Infect. Dis. 4: 6.

Quale, J.M., G.Y. Lipschik and A.E. Heurich. 1987. Management of tuberculous pericarditis. Ann. Thorac. Surg. 43: 653–655.

Riquelme, A., M. Calvo, F. Salech, S. Valderrama, A. Pattillo, M. Arellano et al. 2006. Value of adenosine deaminase (ADA) in ascitic fluid for the diagnosis of tuberculous peritonitis: a meta-analysis. J. Clin. Gastroenterol. 40: 705–710.

Saglam, L., M. Akgun and E. Aktas. 2005. Usefulness of induced sputum and fibreoptic bronchoscopy specimens in the diagnosis of pulmonary tuberculosis. J. Int. Med. Res. 33: 260–265.

Sarkar, S.K., G.S. Sharma, P.R. Gupta and R.K. Sharma. 1980. Fiberoptic bronchoscopy in the diagnosis of pulmonary tuberculosis. Tubercle. 61: 97–99.

Schoch, O.D., P. Rieder, C. Tueller, E. Altpeter, J.P. Zellweger, H.L. Reider et al. 2007. Diagnostic yield of sputum, induced sputum, and bronchoscopy after radiologic tuberculosis screening. Am. J. Respir. Crit. Care. Med. 175: 80–86.

Seibert, A.F., J. Haynes Jr., R. Middleton and J.B. Bass Jr. 1991. Tuberculous pleural effusion. Twenty-year experience. Chest 99: 883–886.

Shakil, A.O., J. Korula, G.C. Kanel, N.G. Murray and T.B. Reynolds. 1996. Diagnostic features of tuberculous peritonitis in the absence and presence of chronic liver disease: A case control study. Am. J. Med. 100: 179–185.

Sherman, S., J.J. Rohwedder, K.P. Ravikrishnan and J.G. Weg. 1980. Tuberculous enteritis and peritonitis. Report of 36 general hospital cases. Arch. Intern. Med. 140: 506–508.

Shingadia, D. and V. Novelli. 2003. Diagnosis and treatment of tuberculosis in children. Lancet Infect. Dis. 3: 624–632.

Simon, H.B., A.J. Weinstein, M.S. Pasternak, M.N. Swartz and L.J. Kunz. 1977. Genitourinary tuberculosis. Clinical features in a general hospital population. Am. J. Med. 63: 410–420.

Singh, M.M., A.N. Bhargava and K.P. Jain. 1969. Tuberculous peritonitis. An evaluation of pathogenetic mechanisms, diagnostic procedures and therapeutic measures. N. Engl. J. Med. 281: 1091–1094.

Sloutsky, A., L.L. Han and B.G. Werner. 2004. Practical strategies for performance optimization of the enhanced Gen-Probe amplified *Mycobacterium tuberculosis* direct test. J. Clin. Microbiol. 42: 1547–1551.

Steingart, K.R., V. Ng, M. Henry, P.C. Hopewell, A. Ramsay, J. Cunningham et al. 2006a. Sputum processing methods to improve the sensitivity of smear microscopy for tuberculosis: A systematic review. Lancet Infect. Dis. 6: 664–674.

Steingart, K.R., M. Henry, V. Ng, P.C. Hopewell, A. Ramsay, J. Cunningham et al. 2006b. Fluorescence versus conventional sputum smear microscopy for tuberculosis: a systematic review. Lancet Infect. Dis. 6: 570–581.

Tuon, F.F., M.N. Litvoc and M.I. Lopes. 2006. Adenosine deaminase and tuberculous pericarditis—a systematic review with meta-analysis. Acta. Trop. 99: 67–74.

Wallace, J.M., A.L. Deutsch, J.H. Harrell and K.M. Moser. 1981. Bronchoscopy and transbronchial biopsy in evaluation of patients with suspected active tuberculosis. Am. J. Med. 70: 1189–1194.

WHO, SEAR. 2015. Tuberculosis control in South East Asian Region. Annual TB Report. WHO.

Woods, G., B.A. Brown-Elliott, P.S. Conville, E.P. Desmond, G.S. Hall, G. Lin et al. 2011. Susceptibility testing of mycobacteria, nocardiae, and other aerobic actinomycetes; approved standard—2nd ed. Clinical and Laboratory Standards Institute (CLSI publication no. M24-A2), Wayne, PA.

World Health Organization. 2015. Global tuberculosis report. WHO, Geneva, Switzerland.

Yajko, D.M., P.S. Nassos, C.A. Sanders, J.J. Madej and W.K. Hadley. 1994. High predictive value of the acid-fast smear for *Mycobacterium tuberculosis* despite the high prevalence of *Mycobacterium avium* complex in respiratory specimens. Clin. Infect. Dis. 19: 334–336.

CHAPTER 6

Microbiological Diagnosis of Viral Diseases

Aashish Choudhary[1] and *Lalit Dar*[2,*]

INTRODUCTION

Viral infections contribute significantly to morbidity and mortality in humans. Besides causing various sporadic, endemic and chronic infections, they are notorious for being the cause of sudden outbreaks and pandemics. Hanging between 'life' and 'non-life', they represent the smallest, simplest and most successful form of parasitism in the entire living world. Diseases caused by viruses range from the common cold (rhinitis) to the lethal Ebola hemorrhagic fever.

The clinical virology laboratory plays a pivotal role in the diagnosis of viral infections (Boivin et al. 2017). Table 1 outlines the approach for laboratory diagnosis in a clinical virology laboratory.

Specimen Collection, Transport and Processing

The first and probably the most crucial step in the laboratory diagnosis of viral infections is an appropriate specimen. The onus of collecting the appropriate specimen lies on the treating clinician taking care of the patient.

[1] Assistant Professor, Virology Section, Department of Microbiology, All India Institute of Medical Sciences, New Delhi, India.
Email: aashishpath@yahoo.co.in

[2] Professor and In-charge, Virology Section, Department of Microbiology, All India Institute of Medical Sciences, New Delhi, India.

* Corresponding author: lalitdar@gmail.com

Table 1. Approaches to the Laboratory Diagnosis of Viral Infections.

Specimen collection, transport and processing	• Serum, plasma, CSF, nasopharyngeal aspirate, stool, urine, etc. • appropriate sample collection and transport to the laboratory
Direct microscopy	• Viral inclusion bodies • Tzanck smear • Electron Microscopy
Antigen detection	• Immunofluorescence test • Enzyme immunoassays (e.g., ELISA) • Immunochromatographic tests (ICT)
Antibody detection (serology)	• ELISA, ICT, etc. • Other tests - Hemagglutination inhibition (HAI) - Complement fixation test (CFT), etc.
Virus isolation	• Cell culture • Embryonated hen eggs • Animal inoculation
Molecular tests/Nucleic acid testing (NAT)	• Conventional DNA polymerase chain reaction (PCR), reverse-transcription PCR, real time PCR • Viral genotype determination • Anti-viral drug resistance testing

An appropriate sample implies the right sample, collected from the correct site and at the correct time. Further, the sample has to be transported under suitable conditions to the laboratory to ensure correct results. 'Garbage in, garbage out' aptly describes the whole process of sample collection, transport and processing. The technique used for collection of a specimen can greatly affect quality of the specimen and, therefore, test results (Dunn 2015).

Some specimens like swabs and tissue/biopsy samples are prone to drying and are better transported in a **viral transport medium** (VTM). The VTM is essentially a buffered salt solution with a protein stabilizer like bovine serum albumin or gelatin, and antibiotics. A commonly used medium is Hank's balanced salt solution (BSS).

Blood: Serum, plasma, whole blood, and some other components like peripheral blood mononuclear cells (PBMCs) have all been used to detect viral pathogens. For antibody detection, **serum** is the usual sample requested. **Plasma** is obtained by centrifuging blood collected in tube containing an anticoagulant. Ethylene diamine tetracetic acid (EDTA) is a commonly used anticoagulant. Heparin is not preferred as it is inhibitory to nucleic acid detection by polymerase chain reaction (PCR). Plasma can be used for nucleic acid testing during the viremic phase of infection. Dried blood spots (DBS) are also increasingly being used for nucleic acid testing owing to the ease of transport and greater flexibility with respect to storage temperature.

Cerebrospinal fluid (CSF) is collected via lumbar puncture in a sterile leak-proof container. No VTM is required for the CSF.

Respiratory samples for viral testing include nasopharyngeal (NP) aspirates, NP swabs, nasal washes and bronchoalveolar lavage (BAL) samples. Nasal and throat swabs are also used for influenza testing with a highly sensitive NAT, as viruses are not a part of the normal flora of the upper respiratory tract, and reflect infection deeper down.

Cutaneous viral lesions like vesicles can be tested for a suspected viral etiology (e.g., HSV 1 and 2, VZV, etc.) by collecting the vesicular fluid using a swab and transporting the latter in VTM. A Tzanck smear can be prepared by scraping the base of such a lesion.

Stool samples are collected in clean, leak proof containers for identification of viruses causing gastroenteritis. Enteroviruses are also detectable in stool specimens since they replicate in the gut.

Tissue and biopsy samples from various sources (lung, intestinal tissue, liver, lymph node, cardiac tissue, brain, etc.) are to be transported in a sterile container in normal saline or VTM. Formalin-fixed paraffin-embedded tissue may be used for nucleic acid testing but considerable deterioration of nucleic acid may have occurred owing to such treatment.

Urine is collected in a sterile container, and is used in the diagnosis of some congenital viral infections [e.g., human cytomegalovirus (HCMV), rubella] and infections in transplant recipients (e.g., BK polyomavirus). Due to acidic pH and high urea content, which can degrade nucleic acids, urine samples should be processed as quickly as possible.

Wherever swabs are being used for viral testing, it is recommended to use **dacron or rayon tipped swabs**. Flocked nylon swabs are increasingly being used since they have better specimen absorption and elution. Cotton and calcium alginate tipped swabs are generally not recommended for viral diagnosis.

Direct Demonstration by Microscopy

The small size of viruses makes direct visualization under a light microscope of limited value in a diagnostic virology laboratory. Viruses can be visualized by negative staining under the electron microscope, but this facility is available only in research and limited number of tertiary care centres.

Although viruses cannot be identified using a light microscope, other evidence of viral infection (e.g., viral inclusion bodies) can be observed under a light microscope. Cytopathological and histopathological techniques often provide direct evidence of viral infections at the site from where these samples are collected (Boivin et al. 2017).

The following are common examples of this diagnostic modality:

- In the **Tzanck test**, scrapings from the base of an unroofed vesicular or ulcerated lesion are smeared onto a slide, air-dried and fixed with 95% alcohol, then stained by the Giemsa or modified Papanicolaou (Pap) stains (Fig. 1). The presence of multinucleate giant cells (with or without intranuclear inclusion bodies) is suggestive of infection with herpes simplex virus types 1 or 2 (HSV-1 or -2) or the varicella-zoster virus (VZV). Inclusion bodies are better visualized in Pap stained smears (Smith and Kubat 2009).
- In transplant recipients on immunosuppressive therapy, and cancer patients on chemotherapy and/or radiotherapy, a common cause of pneumonitis is **HCMV**. In such patients, Giemsa or Pap stained smears from tracheal aspirates, bronchial washings/brushings or BAL fluid, etc. may reveal large intranuclear 'owl eye' inclusion bodies in the nucleus.

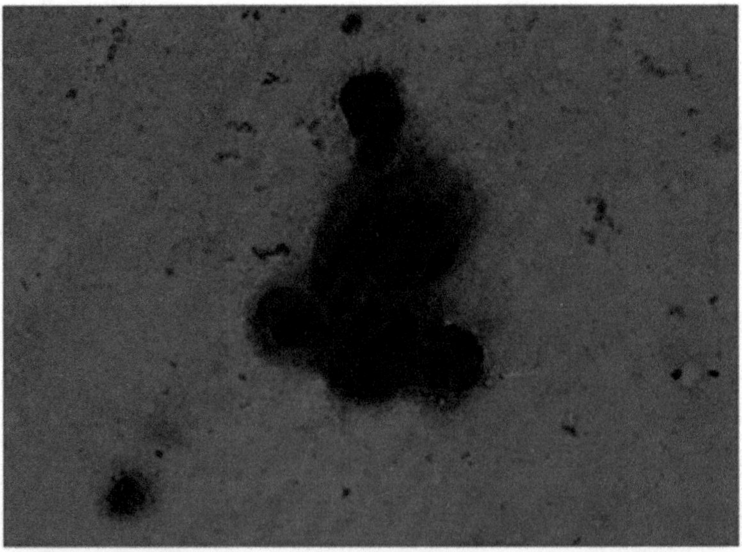

Fig. 1. Tzanck smear Showing Multinucleate Cells with Reniform Moulding of a Few Nuclei, Suggestive of Infection with HSV-1 or -2, or VZV.

- In renal transplant recipients with **BK polyomavirus** (BKPyV) nephropathy, urinary sediment may reveal 'decoy cells' in the urinary sediment.

Although relatively specific for certain viruses, techniques that aim to demonstrate cytopathological or histopathological changes lack in sensitivity. When detected, they provide an important clue to a viral infection in the form of direct microscopic evidence of the offending pathogen.

An important technique in direct microscopy for viral diagnosis is **electron microscopy** (EM). Owing to excellent resolution (~ 0.001 μ) of the EM, direct visualization of the viral pathogen is possible. This may be used for the identification of diarrheagenic viruses in stool and pox viruses in vesicular fluid. Immune electron microscopy (IEM) is immunospecific due to the interaction of a virus with its respective antibody and also more sensitive due to the formation of virus clumps which are easier to detect. The use of colloidal-gold-labelled antibodies further helps identify virus particles or morphological subviral units. Although technically not very demanding, this facility is available only in tertiary care or research institutions.

Antigen Detection

Antigen detection techniques usually target surface antigens or free circulating antigens of the virus in the clinical sample. Although not as sensitive as molecular techniques, these tests are considered more specific than antibody detection based tests since they provide a direct evidence of the existence of the viral pathogen in the sample. This is because cross reactions are not likely as is the case with antibody detection tests.

Various techniques are available to detect viral antigen (Bagher 2010). Many of these techniques can also be adapted for antibody detection.

Immunofluorescence is an immunological test which is based on the detection of antigen by a specific antibody coupled to a fluorescent dye using a fluorescent microscope.

Enzyme immunoassays (EIAs) are solid-phase assays where the antigen is detected by an enzyme labeled antibody combined with colorimetric detection (Leland 2009). Antibodies or antigens present in serum (or other body fluids) are captured by corresponding antigen or antibody coated on to a solid polystyrene surface (usually a 96-well microplate). The enzyme conjugated with the antibody (or antigen) reacts with a colorless substrate-chromogen system to generate a colored product, which indicates the presence of the antigen (or antibody).

Chemiluminescence based immunoassays (CLIAs) is a variety of immunoassay based on antigen antibody reaction only with a difference in respect to the detection system. Luminescence is described as the emission of light from a substance as it returns from an electronically excited state to ground state. When luminescence is produced due to result of a chemical reaction it is termed as chemiluminescence. Antibodies or antigens present are captured by corresponding antigen or antibody coated on to the solid surface. An enzyme conjugated with an antibody reacts with a chemiluminescent substance in turn exciting it by the oxidation and forming intermediates. When the excited intermediates return back to their stable ground state, a photon is released, which is detected by the luminescent signal instrument. Such tests are highly amenable to automation and high throughput testing and are widely used in commercially available antigen/antibody assay systems.

Immunochromatographic tests (ICTs; lateral flow assays) are carried out on chromatographic strips and utilize capillary action. Two kinds of specific antibodies against antigen are used. One of the antibodies is immobilized on the chromatographic paper, and the other is mobile, labeled with colloidal gold and infiltrated into sample pad. The antigen forms an immune complex with the mobile antibody labeled with colloidal gold. The complex moves along with the liquid phase and makes contact with the antibody immobilized on the membrane, followed by formation of an immunocomplex 'sandwich' with it, resulting in a color change visualized as burgundy red, which indicates the presence of the specific antigen in the sample (Leland 2009).

Latex agglutination test is an indirect agglutination test based on antigen antibody reaction where the antigen (or antibody) is coated onto latex microbead particles which, on reaction with the corresponding antibody (or antigen), forms white clumps with background clearing, that can be visualized by the naked eye.

The advantages of antigen detection methods include ease of performance, cost effectiveness, rapid turn-around-time, and a good sensitivity and specificity.

Common applications for diagnosis of viral infections include the detection of HBsAg and HBeAg for hepatitis B, and NS1 antigen for dengue, usually by ELISA, in serum samples. Respiratory viral antigens are detected by immunofluorescence for the diagnosis of influenza, respiratory syncytial virus (RSV) and parainfluenza virus infection. Immunofluorescence is also used for the detection of HCMV pp65 antigen in blood neutrophils. Rotavirus antigen can be detected in feces by ELISA or a latex agglutination test. The latter is a rapid diagnostic test (RDT), i.e., one of the tests which can give results within 30 minutes. ICT-based RDTs are also available for some of these applications and the variety and use of such tests is growing rapidly. These tests are also known as point-of-care (POC) tests, since they are visually readable, do not require any special instruments and can be used under field conditions. The main limitation of RDTs for antigen/antibody detection is that their sensitivity is currently lower than EIAs.

Antibody Detection (Serological) Tests

Serological tests for antibody detection are the most widely used tests in the diagnosis of viral infections. Detection of virus-specific antibody has various applications. Diagnosis of acute or recent infection is based on detection of virus-specific IgM antibody, or the demonstration of a four-fold change in IgG antibody titre compared with a convalescent sample.

Detection of virus-specific IgG antibody indicates past exposure and/or immunity to the concerned virus. IgG antibody is also used for sero-epidemiological studies to ascertain the prevalence of a virus in a population.

Serological tests are the mainstay of diagnosis for some infections, e.g., antibody detection for human immunodeficiency virus (HIV) and hepatitis C virus (HCV), and antibody/antigen detection for hepatitis B virus (HBV), though NATs may subsequently be used for confirmation of infection or for deciding on initiation of therapy and its monitoring.

Another important application of virus-specific IgG is during pre-transplant screening of patients undergoing solid organ transplantation (SOT). For HCMV, which is the most common transplant-associated viral infection, both the donor (D) and recipient (R) are tested for virus-specific IgG and the pair is categorised according to positivity or negativity for the virus-specific IgG, i.e., D+/R+, D+/R-, D-/R+ and D+/R-. Primary infection in the recipient from an infected donor in a D+/R- pair, is a major concern.

During immunosuppression (as induced in transplant recipients) or due to immune deficiency (e.g., in HIV-infected individuals), latent viruses (e.g., those of the *Herpesviridae* family) may reactivate to cause active infection. Reactivation of such latent viruses is usually not associated with a rise in virus-specific IgM. Needless to say, immunocompromised patients may not be able to mount an antibody response to pathogens.

Various techniques are available to detect antibody, and most of these are the same techniques used for antigen detection. The principles of these tests have already been discussed in the previous section. Besides these, other traditional techniques, which have been used for viral diagnosis and epidemiology, include the hemagglutination inhibition (HAI) test, complement fixation test (CFT) and neutralization test. They are used in specific situations and are available in specialized virology laboratories only.

Nucleic Acid Tests (NATs)

Viral nucleic acid tests have totally changed the face of clinical and diagnostic virology. Although many amplification techniques have evolved over the last three decades, polymerase chain reaction is the most popular and commonly used technique in a molecular virology laboratory today (WHO 2011). The sensitivity and specificity of NATs is usually higher than other techniques (including the traditional gold standard, virus culture) (Yang and Rothman 2004).

Polymerase chain reaction (PCR) is a technique for DNA replication that allows a "target" DNA sequence to be selectively amplified. PCR uses the enzyme DNA polymerase that directs the synthesis of DNA from deoxynucleotide substrates on a single-stranded DNA template (Cardenas and Alby 2016). DNA polymerase adds nucleotides to the 3′ end of a custom-designed oligonucleotide, i.e., primer when it is annealed to a longer template DNA. It involves three steps which are denaturation of double stranded DNA followed by annealing of primers and finally extension of the complementary strands. To check whether the PCR successfully generated the amplimer or amplicon, agarose gel electrophoresis is employed for size separation of the PCR products. The size(s) of PCR products is determined by comparison with a DNA ladder, a molecular weight marker which contains DNA fragments of known size (Hoffman and Roshal 2010).

PCR can also be used for RNA viruses, for which the viral RNA is first converted to complementary DNA (cDNA) by using the enzyme **reverse transcriptase**. This variant of PCR is known as the **reverse transcription PCR** (RT-PCR).

Real-time PCR or qPCR is a special type of Polymerase Chain Reaction where both amplification and detection of amplicon is done simultaneously in the same reaction vessel without opening it. Here detection is done by either non-specific fluorescent dyes that intercalate with any double-stranded DNA, or sequence-specific DNA oligonucleotides (probes) that are labelled with a fluorescent dye which permits detection only after hybridization of the probe with its complementary sequence. The software plots the change in fluorescence generated versus the PCR cycle number. The cycle threshold (Ct) is that cycle of the PCR at which the fluorescence in the test exceeds the threshold above background level (Fig. 2). The advantage over conventional PCR are less chance of contamination, quantification of the template and real time monitoring of the PCR reaction (Espy et al. 2006).

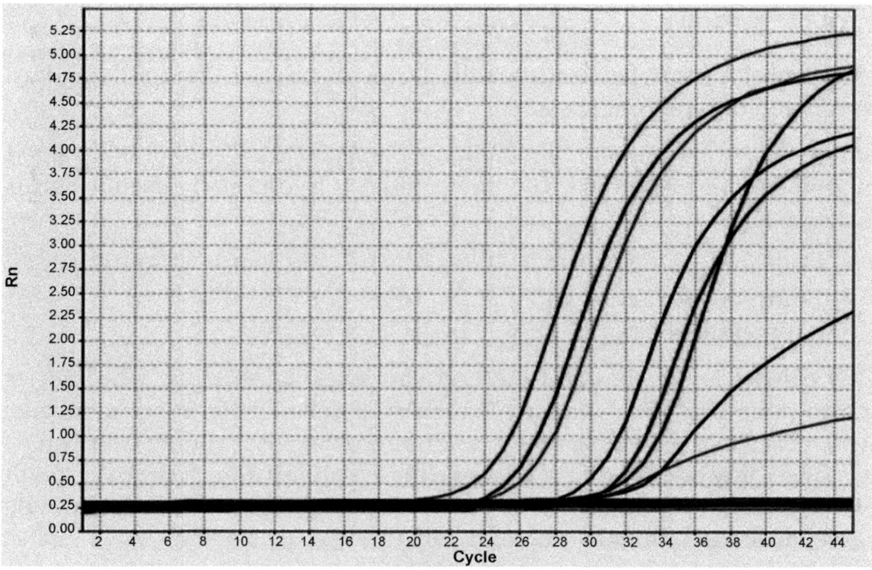

Fig. 2. Real Time PCR Graphs Depicting Amplification Curves with Threshold Cycle (Ct) Plotted Versus Fluorescence Detected.

Viral quantification, i.e., determination of viral load is most commonly performed by the real time PCR technique, and is part of the standard of care for management of chronic/persistent viral infections including HIV-1 and -2 (WHO 2016), HBV (WHO 2015) and HCV infection (WHO 2014). Further, in SOT and hematopoietic stem cell transplant (HSCT) recipients, quantification of HCMV is important in patient management and in deciding pre-emptive therapy (Sloma et al. 2009). Another example is monitoring of viral load of the BK polyomavirus (BKPyV) in renal transplant recipients in the urine and plasma (Ambalathingal et al. 2017).

Besides NATs, molecular tests (often followed by sequencing) are also used for genotyping and detection of anti-viral drug resistance associated mutations. Novel and emerging viruses are usually detected and identified by NATs and sequencing, since commercial antigen/antibody tests are not available initially and virus cultivation may either not be possible or may have to be attempted on a large variety of cell lines before identifying a suitable one. Starting with the hepatitis C virus, molecular techniques have proved extremely useful for many such viral infections and for viral outbreaks like those due to the newer coronaviruses (severe acute respiratory syndrome virus, SARS; Middle-East respiratory syndrome virus, MERS) and other emerging viruses like Nipah virus, Zika virus, etc. (Petersen 2016).

Virus Isolation

Isolation of a virus has traditionally been considered the gold standard technique for viral diagnosis.

Since viruses are obligate intracellular parasites, they cannot be isolated in artificial cell-free media. The three methods traditionally used for virus isolation are:

- Cell culture
- Embryonated hen eggs
- Animal inoculation

Cell culture is the standard method employed by virology laboratories which attempt virus isolation. A variety of cells (Table 2) are available, which vary in their susceptibility to different viruses. Primary cells can be passaged only once or twice, but have the capability to support the growth of a wide range of viruses. On the other hand, continuous cell lines can be sub-cultured indefinitely (Schmidt 1979).

In general, virus isolation in cell culture takes few days to weeks (the latter for slow growing viruses like HCMV). This limitation was circumvented by the development of shell vial (centrifugation enhanced) cultures in which immediate early antigens can be detected, with results available in about 48 hours.

Table 2. Cell Culture for Virus Isolation.

Cell culture type	Description	Karyotype (Ploidy)	Examples	Number of passages (sub-cultures)
Primary cells	Derived from tissue, and have been propagated *in vitro* for the first time	Same chromosome number as parental tissue (diploid)	Monkey kidney, chick embryo cells	1–2
Diploid (semi-continuous) cell lines	Develop during subculture of primary cells	75–100% cells have the same karyotype as cells from the species of origin (diploid)	Human embryo lung fibroblasts, e.g., MRC-5 and WI-38 cell lines	20–50
Continuous (heteroploid) cell lines	Have the ability of indefinite passages due to cell immortalization; commonly derived from cancer cells	Exist as a population of cells with $< 75\%$ of the population having a diploid chromosome constitution	Human cell lines like HEp-2, HeLa; mammalian cell lines like Vero	Indefinite

Many viruses produce morphological changes (cytopathic effect, CPE) in the cells in which they replicate (Fig. 3). These are often typical of a given virus in a given cell line to make a presumptive diagnosis. Viruses which do not induce a CPE can be detected by immunofluorescence, hemadsorption, interference, molecular methods, etc.

Cell culture has advantages which include relatively good sensitivity for virus isolation, potential to detect many different viruses if sufficient cell lines are used, and the ability to provide a viral isolate which is available for further molecular characterization. Limitations of the technique include slow-turn-around-time, requirement of technical expertise, inability of certain viruses to grow in cell culture, etc.

Genetically engineered cell lines are now available which allow the isolation of viruses in cell lines not ordinarily permissive to that virus.

Embryonated hen eggs were commonly used for virus isolation till the advent of cell culture. The milieu provided by the embryonated egg supports the growth of many viruses infecting humans. Various routes are available for virus cultivation, including the amniotic cavity (primary isolation of influenza virus), allantoic cavity, yolk sac (arboviruses like the yellow fever vaccine strain 17-D) and chorioallantoic membrane (CAM; cultivation of pox viruses and HSV). Currently, the major use of this technique is in the mass scale production of seasonal influenza vaccines using the allantoic route.

Laboratory animals were commonly used for virus isolation till advancement in cell culture made this method unnecessary for routine use. Laboratory animals like suckling and adult mice were commonly used for viral detection. Nevertheless, experimental animal are still required for viruses which fail to grow in embryonated eggs and cell culture.

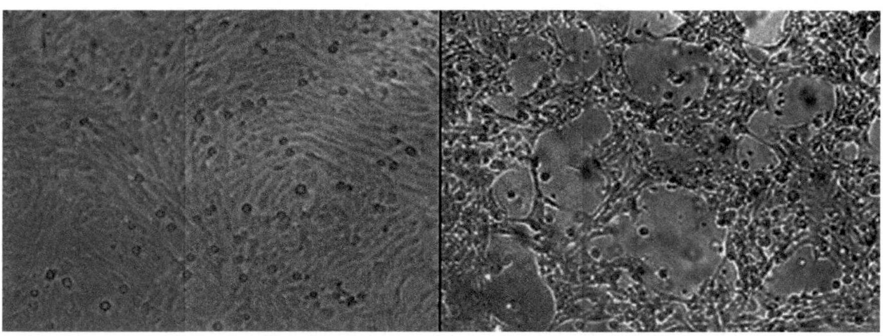

Fig. 3. Uninoculated Madin-Darby Canine Kidney (MDCK) Cells [Left], and the 'Lacing' Pattern CPE of Influenza Virus [Right].

Advancements in molecular techniques have significantly reduced the use of embryonated hen eggs and laboratory animals for virus isolation in a diagnostic laboratory.

References

Ambalathingal, G.R., R.S. Francis, M.J. Smyth, C. Smith and R. Khanna. 2017. BK Polyomavirus: Clinical aspects, immune regulation, and emerging therapies. Clin. Microbiol. Rev. 2017 Apr; 30(2): 503–528.

Bagher, F. 2010. Diagnosis by viral antigen detection. pp. 113–132. *In*: Jerome Keith, R. (ed.). Lennette's Laboratory Diagnosis of Viral Infections, Fourth Edition. Informa Healthcare, New York, NY.

Boivin, G., T. Mazzulli and M. Petric. 2017. Diagnosis of viral infections. pp. 291–320. *In*: Richman, D.D., R.J. Whitley and F.G. Hayden (eds.). Clinical Virology, Fourth Edition. ASM Press, Washington, DC.

Cardenas, A.M. and K. Alby. 2016. Nucleic acid amplification by polymerase chain reaction. pp. 129–136. *In*: Loeffelholz, M.J., R.L. Hodinka, S.A. Young and B.A. Pinsky (eds.). Clinical Virology Manual, Fifth Edition. ASM Press, Washington, DC.

Dunn, J.J. 2015. Specimen collection, transport, and processing: Virology. pp. 1405–1421. *In*: Jorgensen, J.H., M.A. Pfaller, K.C. Carroll, G. Funke, M.L. Landry, S.S. Richter et al. (eds.). Manual of Clinical Microbiology, Eleventh Edition. ASM Press, Washington, DC.

Espy, M.J., J.R. Uhl, L.M. Sloan, S.P. Buckwalter, M.F. Jones, E.A. Vetter et al. 2006. Real time PCR in clinical microbiology: Applications for routine laboratory testing. Clin. Microbiol. Rev. 2006 Jan; 19(1): 165–256.

Hoffman, N. and M. Roshal. 2010. Design of molecular virologic tests. pp. 59–73. *In*: Jerome Keith, R. (ed.). Lennette's Laboratory Diagnosis of Viral Infections, Fourth Edition. Informa Healthcare, New York, NY.

Leland, D.S. 2009. Enzyme immunoassays and immunochromatography. pp. 89–102. *In*: Specter, S., R.L. Hodinka, S.A. Young and D.L. Wiedbrauk (eds.). Clinical Virology Manual, Fourth Edition. ASM Press, Washington, DC.

Petersen, E.E. 2016. Interim Guidelines for Pregnant Women During a Zika Virus Outbreak – United States, 2016. MMWR Morb. Mortal Wkly. Rep. [Internet] [cited 2017 Apr 19]; 65. Available from: http://www.cdc.gov/mmwr/volumes/65/wr/mm6502e1.htm.

Schmidt, N.J. 1979. Cell culture techniques for diagnostic virology. pp. 65–139. *In*: Lennette, E.H. and N.J. Schmidt (eds.). Viral, Rickettsial and Chlamydial Infections, Fifth Edition. American Public Health Association, Inc., Washington, DC.

Sloma, C.R., T.E. Grys and R.R. Razonable. 2009. Cytomegalovirus. pp. 69–90. *In*: Hayden, R.T., K.C. Carroll, Tang Yi-Wei, D.M. Wolk (eds.). Diagnostic Microbiology of the Immunocompromised Host. ASM Press, Washington, DC.

Smith, R.D. and A. Kubat. 2009. The cytopathology of virus infection. pp. 52–63. *In*: Specter, S., R.L. Hodinka, S.A. Young and D.L. Wiedbrauk (eds.). Clinical Virology Manual, Fourth Edition. ASM Press, Washington, DC.

WHO, Asia RO for S-E. 2011. Establishment of PCR laboratory in developing countries [cited 2017 Apr 19]; Available from: http://www.who.int/iris/handle/10665/205020.

WHO. 2014. Guidelines for the screening, care and treatment of persons with hepatitis C infection [Internet]. WHO [cited 2017 Apr 21]. Available from: http://www.who.int/hepatitis/publications/hepatitis-c-guidelines/en/.

WHO. 2015. Guidelines for the prevention, care and treatment of persons with chronic hepatitis B infection [Internet]. WHO [cited 2017 Apr 21]. Available from: http://www.who.int/hepatitis/publications/hepatitis-b-guidelines/en/.

WHO. 2016. Consolidated guidelines on HIV prevention, diagnosis, treatment and care for key populations [Internet]. WHO [cited 2017 Apr 21]. Available from: http://www.who.int/hiv/pub/guidelines/keypopulations-2016/en/.

Yang, S. and R.E. Rothman. 2004. PCR-based diagnostics for infectious diseases: uses, limitations, and future applications in acute-care settings. Lancet Infect. Dis. 2004 Jun; 4(6): 337–48.

CHAPTER 7

Microbiological Diagnosis of Fungal Infections

Gagandeep Singh[1,*] and *Immaculata Xess*[2]

INTRODUCTION

Invasive fungal infections (IFI) are associated with high morbidity, mortality and healthcare costs especially in the immunocompromized hosts (Marchetti et al. 2010, Zaragoza and Pemán 2008). In recent surveys, a rise in incidence of IFIs have also been reported in immunocompetent critically ill patients though the diagnosis of IFIs is difficult to ascertain in such patients due to the lack of definitions and inaccurate diagnostic procedures (Meersseman et al. 2007). Due to poor sensitivity of conventional diagnostic tests, the European Organization for Research and Treatment of Cancer/Invasive Fungal Infections Cooperative Group and the National Institute of Allergy and Infectious Diseases Mycoses Study Group (EORTC/MSG) recommended classification of the IFIs in immunocompromised hosts into proven, probable and possible fungal infection depending on host, clinical and mycological factors (De Pauw et al. 2008). However, there is no such recommendation available for immunocompetent critically ill patients. For these reasons, attempts are being made for new laboratory tests, and better management strategies by stratification of patients in critical care setting. This would help in early therapeutic interventions of high risk

[1] Assistant Professor, Mycology Section, Department of Microbiology, All India Institute of Medical Sciences, Ansari Nagar, New Delhi-110029, India.
[2] Professor, Mycology Section, Department of Microbiology, All India Institute of Medical Sciences, Ansari Nagar, New Delhi-110029, India.
* Corresponding author: drgagandeep@gmail.com

patients to reduce the high mortality rate associated with IFIs. Antifungal pharmacotherapy has also seen important developments in recent times. Management of fungal infections is complicated by drug interaction and toxicity of antifungal drugs. This chapter addresses the diagnosis of fungal infections along with newer approaches.

Fungi – Common Classification

In clinical practice, fungi are classified into three main groups; the yeast and yeast like, the molds, and the dimorphic fungi. Yeasts are unicellular organisms which multiply by budding. The two important pathogenic yeasts include *Candida* and *Cryptococcus*. Other less often encountered yeast species are *Trichosporon, Geotrichum, Rhodotorula*, etc. which can cause infections in immunocompromised hosts. The molds are long filamentous branching organisms which may be septate (e.g., *Aspergillus, Fusarium, Scedosporium*, etc.) or aseptate/pauciseptate (Mucormycetes). The molds can also be classified into two broad groups, the phaeohyphomycetes and the hyalohyphomycetes, based on color of the fungi. The phaeohyphomycetes are black and hyalohyphomycetes are hyaline. The dimorphic fungi have two morphologies and occur as yeast form in tissue and mold form in the environment. The dimorphic fungi include *Histoplasma capsulatum, Blastomyces dermatitidis, Coccidioides immitis, Paracoccidioides brasiliensis, Sporothrix schenckii* and *Talaromyces marneffei*.

Diagnosis of Fungal Infections: If You Don't Suspect it, You'll Miss it!

The present status of diagnosis of IFIs can be summed up as "too little and too late". The symptoms of IFIs are typically non-specific and not unlike those of other infections. Although clinical signs of fungal infections are mostly non-descript relative to the infecting pathogen, there are few pathogen-specific clinical presentations, which may serve as clues for diagnosis. Specific signs on imaging may also help in diagnosis. Table 1 summarizes the various clinical and radiological signs which can help the clinician in suspecting IFIs.

Conventional diagnosis of these infections, based on direct demonstration and culture of the offending organism, is often insensitive and takes time. Patients who have been on broad spectrum antibiotics for extended periods of time without improvement in systemic signs of infection or 'sepsis' are likely to have fungal infections. Certain groups of patients who have specific predisposing factors that favor infections due to fungi, e.g., patients on total parenteral nutrition (TPN) are at risk of fungemia with *Candida*

Table 1. Clinical and Radiological Signs in Commonly Occurring IFIs.

Invasive Candidiasis
• Weight loss
• Abdominal pain
• Prolonged antibiotic-resistant fever
• Hepatic and/or splenic enlargement
• Small radiolucent lesions in the liver or spleen (CT scan)
• Findings of endophthalmitis on fundoscopy
• Non-tender, popular/nodular, erythematous skin lesions
Invasive Aspergillosis
• Prolonged antibiotic-resistant fever
• Haemoptysis (aspergilloma and chronic necrotizing pulmonary aspergillosis)
• Cough and hypoxia
• Pleuritic pain during neutropenic fever
• Acute tracheobronchitis, bronchiolitis, and bronchopneumonia
• Halo or air-crescent sign (CT scan)
Invasive Mucormycosis
• Sinus involvement in addition to pulmonary disease
• Reverse halo sign (CT scan)

and *Malassezia.* Immunocompromised patients admitted to hospitals with construction activity are at increased risk for invasive aspergillosis.

Sample collection, transport and processing

Accurate diagnosis of fungal diseases is highly dependent on the appropriate sample. In addition, it is important for the clinicians to convey the related diagnostic possibilities to the laboratory based on clinical presentation as processing of samples may be different for different diagnoses.

For invasive infections tissue biopsy or a fine needle aspirate is the preferred sample. However, many of the patients with IFIs also have thrombocytopenia, making profuse bleeding a real possibility during the deep tissue collection. The tissue biopsy should be sent immediately to the mycology lab in a sterile container with sterile saline to prevent drying of the sample which can be detrimental to the recovery of the fungus. The differential diagnosis holds key to the type of processing the tissue sample will undergo in the laboratory. For a suspected case of mucormycosis the tissue sample should be cut into small pieces and allowed to digest. Whereas in all the other cases of IFIs, including invasive aspergillosis, the tissue should be homogenized by using a mechanical or hand held tissue homogenizer or a mortar and pestle. In addition to the cutaneous and subcutaneous mycoses, the skin lesions can be a part of disseminated fungal infections. Therefore, appropriate skin biopsy sent in time may help in clenching the diagnosis without the need for a more invasive procedure. Skin biopsy is processed as any other tissue. Fungi often cause deep seated infections. Pus samples from ultrasound/CT guided fine needle aspiration from the abscesses should be sent immediately in a sterile container.

For the blood culture the procedure is like routine bacterial blood culture, however, larger blood volume may yield better results. The load of pathogenic fungi is usually low in respiratory samples. Therefore, these samples should be concentrated to enhance recovery. *Candida* sp. are usually considered as colonizers until and unless there is demonstration of yeast in a lung biopsy. Similarly, *Candida* sp. are also considered colonizers of the urinary tract unless the patient is neutropenic, is a pre-term infant or is to undergo any kind of urologic procedure. The dermatophytes are resistant to drying and therefore samples like skin scrapings, hair and nail clippings can be collected in black thick paper sachets and transported in a regular envelope.

In patients with suspected fungal keratitis, the sample should either be inoculated by the clinician bedside on a blood agar plate or the scrapings can be sent between two sterile glass slides, one for direct examination and the other for culture. The vitreous tap and aqueous tap samples should be transported in the syringe only.

Samples from patients with suspected fungal infections should not be collected using swabs as the samples dry quickly, leading to non-viability of fungi. Only in rare cases where there are superficial ulcers or aural lesions, moist swabs may be used for collection of pus samples.

Conventional methods for diagnosis of fungal infections

The conventional diagnosis of fungal infections is based on the direct demonstration of the fungal elements in the tissue specimen and culture of the fungus.

Direct demonstration

The most common method used for the direct demonstration is the 10% potassium hydroxide (KOH) mount with calcofluor white. The KOH-calcofluor white is a highly sensitive technique and an essential tool in any mycology laboratory. However, the correct reporting of fungal elements needs experience. In cases where the fungal burden is low the test may be negative. The morphology seen on the direct demonstration may give a clue about the pathogen and therefore be of use in deciding the correct antifungal agent. The agents of mucormycosis are thin walled, aseptate or pauci-septate, may show ribbon like folding and right angle branching. The septate hyphae which show dichotomous branching and division at 45° may be indicative of aspergillosis, fusariosis, sceodosporiosis, etc. Certain dematiaceous fungi do not get stained well with calcofluor white and therefore in clinical cases with high suspicion of one, e.g., chromoblastomycosis, brain abscess, etc. the sample should also be examined under a light microscope. The India ink stain can be used to demonstrate capsulated yeasts in cerebrospinal fluid which can suggest presence of *Cryptococcus neoformans*. *Pneumocystis jiroveci* can be demonstrated in respiratory samples by Gomori methenamine silver stain, Toluidine blue O stain or on direct immunofluorescence. Yeast forms of the dimorphic fungi can be demonstrated by performing Giemsa staining on impression smears of tissue sections or bone marrow.

Culture

Although still considered the gold standard for the diagnosis of fungal infections, culture is not a reliable diagnostic tool and has a long turnaround time. Blood cultures for patients with candidemia may be negative in up to 50% of the patients with disseminated candididasis. The percent positivity of blood culture may be even less in other yeasts and molds like *Fusarium* spp. and *Scedosporium* spp. As fungi are ubiquitous, positive fungal cultures should always be interpreted with caution and correlated with clinical picture and direct microscopic findings. Where in doubt, always ask the clinician to send a repeat sample for definitive diagnosis. The identification of yeasts can be performed using automated systems such as the Vitek-2 system, etc. The MALDI-TOF (matrix-assisted laser desorption/ionization/time of flight) can be used for the identification of yeasts and molds obtained after culture. The protein content must be extracted using various protocols involving mechanical disruption, acetonitrile and formic acid. The turnaround time with this procedure is much shorter.

Antifungal susceptibility testing

Although the broth microdilution method by the CLSI is the recommended method for the susceptibility testing of yeasts and molds, it is expensive, demands technical expertise and is labor intensive. Automated systems can be used to perform susceptibility testing of yeasts. There are a few centers which use the disk diffusion method/E-test.

Non-culture based diagnosis

There are several non-culture based tests which may improve the diagnostic accuracy of IFIs in the appropriate clinical settings.

A. Procalcitonin

Till date, no surrogate marker is identified to clearly indicate invasive fungal infection. Martini et al. (Martini et al. 2010) had studied the role of procalcitonin in the diagnosis of candidemia. They found that the level of PCT in patients with fungaemia is lower than those with bacteremia. Therefore, they suggested that in the scenario of a consistent clinical picture, low level of PCT may indicate candidemia.

B. Cryptococcal capsular polysaccharide antigen

There are several commercially available kits for the detection of cryptococcal capsular polysaccharide antigen in clinical samples especially cerebrospinal fluid and serum. The sensitivity and specificity is high. The detection can be done by latex agglutination, dip stick assay as well as lateral flow format. Most of these can be performed on the patient's bedside and have been used as point-of-care tests. A negative test does not exclude cryptococcal infection. False-negative reactions may be caused by low titers, early infection, prozone effect or poorly encapsulated organism. False-positive reactions can occur due to the presence of rheumatoid factor, agar syneresis fluid, *Capnocytophaga animorsus*, *Trichosporon* sp., improper cleaning of the ring slide, hydroxyethyl starch, etc. Pronase treatment has been shown to reduce false positives and increase sensitivity in serum specimens. A positive test is quantitated and followed for disease activity and response to treatment.

C. β-D-glucan (BG) assay

The (1→3)-β-D-glucan (BDG) is present in the cell wall of almost all fungi except *Cryptococcus* sp. and mucormycetes. In patients with hematologic malignancies who had definitive or suspected invasive fungal disease, the test showed a specificity of 62.4% and sensitivity of 92.4% (Ostrosky-Zeichner 2004). A positive test may be useful in initiating pre-emptive antifungal treatment in the immunosuppressed patients. High levels of BDG have also been observed in patients with *Pneumocystis jiroveci* pneumonia. False-positive results have been seen in patients on albumin, immunoglobulins, those undergoing hemodialysis using cellulose membranes, etc. Therefore, a single positive test may have less diagnostic value. Serial testing might be more predictive of true infection. The performance of the BDG assay in immunocompetent patients is unknown, especially in critically ill patients in the ICU.

D. Galactomannan antigen test

Galactomannan (GM) is a cell wall component of *Aspergillus*. Detection of galactomannan in serum and BAL fluid has been studied in patients with neutropenia and/or hematologic malignancy. Galactomannan is reported as an index of optical density (galactomannan index [GMI]; the index is considered to be positive when 2 aliquots from the same sample have an optical density > 0.5) (Miceli et al. 2008). The GM test is standardized, widely available and reproducible (Wheat and Walsh 2008). It has been accepted by regulatory agencies as a surrogate marker for the diagnosis of invasive aspergillosis in trials (Cornely et al. 2007, Ullmann et al. 2007). Serum GM testing can be used as a surrogate end point for aspergillosis outcome in patients with hematological malignancy (Anaissie 2007). Though earlier studies showed piperacillin/tazobactam antibiotic use may provide false positive reaction, recent kits have overcome this problem to a certain extent. Enteral feeding with soybean protein, gastrointestinal colonization with *Bifidobacterium* in neonates can also result in false-positives. Combining the BAL galactomannan assay with PCR may increase the detection rate of invasive aspergillosis in neutropenic patients, though PCR techniques are in the process of standardization.

A common diagnostic predicament occurs in the ICU when *Aspergillus* is isolated from a respiratory sample from patients lacking the traditional risk factors for invasive disease. This finding could indicate invasive infection or mere colonization or could be a marker for the development of future invasive disease. The predictive value of *Aspergillus* isolation from lower respiratory tract sample as a marker for development of invasive disease is high in immunocompromised patients such as those with HSCT and neutropenia. Isolation of *Aspergillus* from respiratory specimens of ICU patients should prompt further diagnostic work-up, even when risk factors are absent.

E. Candida mannan and anti-mannan antibody assays

The detection of mannan and anti-mannan antibody has been used in immunocompromised patient with hematological malignancy and those who have undergone abdominal surgery. The sensitivity and specificity is poor when used alone. However, the combined usage increases the sensitivity to 85%. More studies are required to ascertain the utility of this assay (Verduyn Lunel et al. 2009, Ellis et al. 2009).

F. Serology (antigen/antibody detection) for endemic fungi

There are two tests available for the detection of fungal antibodies in serum. Immunodiffusion is specific although it lacks sensitivity. The complement fixation test is sensitive but lacks specificity. Therefore, it is advised that these two test should be used in combination. More recent addition is the detection of the *Histoplasma* polysaccharide antigen in serum and body fluids especially urine by ELISA. The sensitivity and specificity of antigen detection in urine is > 90% of each (Theel et al. 2015, Theel et al. 2013). A similar test is now being used for blastomycosis.

G. Molecular tests

Though nucleic acid amplification techniques can detect fungal DNA in patient samples and PCR assays hold promise for diagnosis of invasive mycoses, standardization and validation is still awaited. Combination of one of the biomarkers to a PCR based assay has been shown to have greater sensitivity and specificity. The European *Aspergillus* PCR initiative (EAPCRI) working group has developed protocols for performing the *Aspergillus* PCR assay. The pooled sensitivity and specificity of this assay in BAL samples is > 90% in cases of invasive aspergillosis (White et al. 2015, Arvanitis et al. 2014).

The T2 Candida is an FDA approved system which uses a ground breaking T2 Magnetic Resonance technology (T2MR). This is the first technology capable of detecting blood stream infections without the need for culture. It has 91.1% sensitivity and 99.4% specificity. The limit of detection is as low as 1 CFU/mL. Species-specific results become available in 4–5 hours directly from whole blood. This assay is highly accurate even in the presence of antimicrobials in the blood (Hamula et al. 2016, Mylonakis et al. 2015).

Although the results of non-culture based tests are promising, the extent to which they affect patient management, clinical outcome and cost-effectiveness of care remains to be determined. Figure 1 shows algorithm for the diagnosis of common invasive fungal infections.

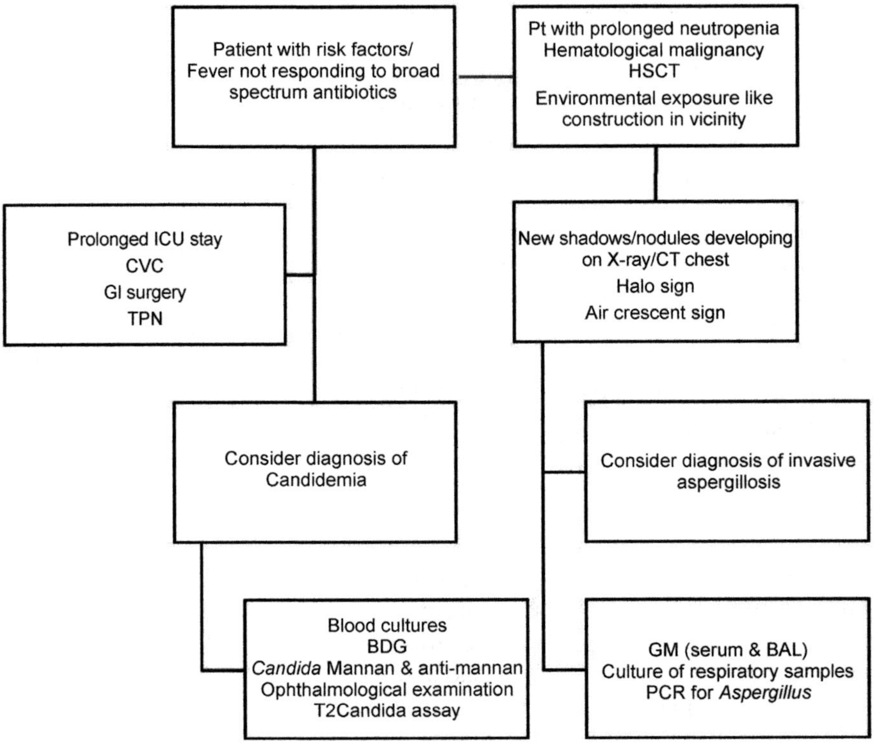

Fig. 1. Algorithm for the Diagnosis of Common Invasive Fungal Infections.

References

Anaissie, E.J. 2007. Trial design for mold-active agents: Time to break the mold—aspergillosis in neutropenic adults. Clin. Infect. Dis. 44: 1298–1306.

Arvanitis, M., P.D. Ziakas, I.M. Zacharioudakis, F.N. Zervou, A.M. Caliendo and E. Mylonakis. 2014. PCR in diagnosis of invasive aspergillosis: A meta-analysis of diagnostic performance. J. Clin. Microbiol. 52: 3731–3742.

Cornely, O.A., J. Maertens, D.J. Winston, J. Perfect, A.J. Ullmann, T.J. Walsh et al. 2007. Posaconazole vs. fluconazole or itraconazole prophylaxis in patients with neutropenia. N. Engl. J. Med. 356: 348–359.

De Pauw, B., T.J. Walsh, J.P. Donnelly, D.A. Stevens, J.E. Edwards, T. Calandra et al. 2008. Revised definitions of invasive fungal disease from the European Organization for Research and Treatment of Cancer/Invasive Fungal Infections Cooperative Group and the National Institute of Allergy and Infectious Diseases Mycoses Study Group (EORTC/MSG) Consensus Group. Clin. Infect. Dis. 46: 1813–1821.

Ellis, M., B. Al-Ramadi, R. Bernsen, J. Kristensen, H. Alizadeh and U. Hedstrom. 2009. Prospective evaluation of mannan and anti-mannan antibodies for diagnosis of invasive Candida infections in patients with neutropenic fever. J. Med. Microbiol. 58: 606–615.

Hamula, C.L., K. Hughes, B.T. Fisher, T.E. Zaoutis, I.R. Singh and A. Velegraki. 2016. T2 Candida provides rapid and accurate species identification in pediatric cases of Candidemia. Am. J. Clin. Pathol. 145: 858–861.

Marchetti, O., P. Eggimann and T. Calandra. 2010. Invasive candidiasis in critically ill patients: does progressing knowledge improve clinical management and outcome? Curr. Opin. Crit. Care. 16: 442–444.

Martini, A., L. Gottin, N. Menestrina, V. Schweiger, D. Simion and J.L. Vincent. 2010. Procalcitonin levels in surgical patients at risk of candidemia. J. Infect. 60: 425–430.

Meersseman, W., K. Lagrou, J. Maertens and E. Van Wijngaerden. 2007. Invasive aspergillosis in the intensive care unit. Clin. Infect. Dis. 45: 205–216.

Miceli, M.H., M.L. Grazziutti, G. Woods, W. Zhao, M.H. Kocoglu, B. Barlogie et al. 2008. Strong correlation between serum aspergillus galactomannan index and outcome of aspergillosis in patients with hematological cancer: clinical and research implications. Clin. Infect. Dis. 46: 1412–1422.

Mylonakis, E., C.J. Clancy, L. Ostrosky-Zeichner, K.W. Garey, G.J. Alangaden, J.A. Vazquez et al. 2015. T2 magnetic resonance assay for the rapid diagnosis of candidemia in whole blood: a clinical trial. Clin. Infect. Dis. 60: 892–899.

Ostrosky-Zeichner, L. 2004. Prophylaxis and treatment of invasive candidiasis in the intensive care setting. Eur. J. Clin. Microbiol. Infect. Dis. 23: 739–744.

Theel, E.S., J.A. Harring, A.S. Dababneh, L.O. Rollins, J.E. Bestrom and D.J. Jespersen. 2015. Reevaluation of commercial reagents for detection of Histoplasma capsulatum antigen in urine. J. Clin. Microbiol. 53: 1198–1203.

Theel, E.S., D.J. Jespersen, J. Harring, J. Mandrekar and M.J. Binnicker. 2013. Evaluation of an enzyme immunoassay for detection of Histoplasma capsulatum antigen from urine specimens. J. Clin. Microbiol. 51: 3555–3559.

Ullmann, A.J., J.H. Lipton, D.H. Vesole, P. Chandrasekar, A. Langston, S.R. Tarantolo et al. 2007. Posaconazole or fluconazole for prophylaxis in severe graft-versus-host disease. N. Engl. J. Med. 356: 335–347.

Verduyn Lunel, F.M., J.P. Donnelly, H.A. van der Lee, N.M. Blijlevens and P.E. Verweij. 2009. Circulating Candida-specific anti-mannan antibodies precede invasive candidiasis in patients undergoing myelo-ablative chemotherapy. Clin. Microbiol. Infect. 15: 380–386.

Wheat, L.J. and T.J. Walsh. 2008. Diagnosis of invasive aspergillosis by galactomannan antigenemia detection using an enzyme immunoassay. Eur. J. Clin. Microbiol. Infect. Dis. 27: 245–251.

White, P.L., C. Mengoli, S. Bretagne, M. Cuenca-Estrella, N. Finnstrom, L. Klingspor et al. 2015. Evaluation of Aspergillus PCR protocols for testing serum specimens. J. Clin. Microbiol. 49: 3842–3848.

Zaragoza, R. and J. Pemán. 2008. The diagnostic and therapeutic approach to fungal infections in critical care settings. Adv. Sepsis 6: 90–98.

CHAPTER 8

Microbiological Diagnosis of Parasitic Diseases

*Nishant Verma** and *B.R. Mirdha*

INTRODUCTION

Medical Parasitology deals with those infections in humans, which are caused by parasites. Diagnosis of the human parasitic infections is a highly challenging field and demands a great deal of experience in the area along with excellent examination and interpretative skills. In contrast to other branches of medical microbiology, there is limited availability of automated procedures for the diagnosis of parasitic diseases, hence, manual techniques for the direct demonstration of causative parasitic organisms are usually the mainstay of diagnosis. Making a correct diagnosis requires a holistic approach and is largely dependent upon a person's skills to appreciate morphological characteristics of various stages of parasitic agents. Additionally, a good knowledge about the parasite, including its life-cycle, geographical distribution, epidemiology, seasonal variations of the disease, mode of transmission, pathogenesis, clinical presentations, and types of susceptible hosts are essential components to arrive at a definitive diagnosis. Further, a good continued interaction between the clinicians and the laboratory personnel is of prime importance to ensure both accurate and timely diagnosis, that improves patient care and management. An informed clinician who provides appropriate clinical samples, in recommended

Parasitology Section, Department of Microbiology, All India Institute of Medical Sciences, New Delhi-110029, India.

* Corresponding author: drnishant2k@yahoo.co.in

amount, in proper container, under optimum transport conditions within the desirable time frame contributes immensely to facilitating a correct diagnosis and interpretation of results. At the same time, knowledge of patient's history along with relevant physical findings and other relevant tests results are of immense importance for a medical/clinical parasitologist to interpret the findings in right perspective. Immuno-deficiency states, extremes of ages, and travel-associated infections are some of the other special circumstances, which may make diagnoses of these infections more complex, possibly due to atypical clinical presentation or the patient may present some rare infections which are not normally prevalent in a particular geographical area.

Considering that a number of infectious and non-infectious diseases may be with similar signs and symptoms, and many of the routine diagnostic investigations may not be specific at distinguishing these, a recommended parasitological investigative work-up, performed well in time, can mean a difference of life and death for a patient (Clinical and Laboratory Standards Institute 2005, Garcia 2009, McVicar and Suen 1994). Depending on the systems or sites involved in an infected/diseased individual, different clinical specimens are desirable in order to confirm the diagnosis. For parasitological work up, several recommended guidelines are available, which have to be strictly adhered to for performing the necessary relevant investigations. This is important to ensure quality of patient care services as well as for proper patient management (Isenberg 1995, 2004, Garcia 2010, Garcia et al. 2003, 2008, Committee on Education, American Society of Parasitologists 1977, College of American Pathologists 2012).

Parasitic Infections of The Intestinal Tract

Fecal specimens

For suspected intestinal tract infections, the fecal sample is the desired clinical sample and fecal examination is performed as early as possible. Examination of fecal samples can be broadly divided into macroscopic and microscopic examinations. As a general rule, fresh fecal specimen is recommended. Substances, such as barium, antimicrobials, charcoal, bismuth, and anti-diarrheal agents, may interfere with parasitological investigations, and the effect may remain for a period of up to 2 wk (Clinical and Laboratory Standards Institute 2005, Garcia 2009, Garcia and Voge 1980, Cartwright 1999, Isenberg 2004, Melvin and Brooke 1982, Sawitz and Faust 1942, Jones and Lopez 2004).

Instructions for patients

For the collection of fecal specimen, patient should be advised to collect the specimen in a clean, wide-mouthed, screw-capped container. It is important to instruct the patient to avoid mixing of urine or water with the fecal specimen. This is because the acidic pH of urine can inhibit the motility of the trophozoites and water may contain some free-living organisms, which may confound as human parasites.

Steps after collection

The specimen container should be kept in a sealed plastic bag, and it is important that all specimens are considered as potential biohazards and are handled carefully. All specimens must be labeled properly, mentioning patient's name, date and time of collection, and patient's unique identification number. With each specimen, a properly filled requisition form must be provided mentioning the relevant patient information and the required investigations.

Number of specimens and frequency of collection

In general, it is recommended that three fecal specimens be collected and sent for parasitological examinations. The specimens should be collected on separate days, preferably every alternate day. However, all the specimens should have been collected and sent within 10 days of the first specimen collection. In cases where there is a suspicion of intestinal amebiasis, six specimens may be recommended, which are to be collected within a 14 day period (Garcia 2016).

It is important that fecal examination is performed as soon as possible after the collection of the specimen. This is particularly so in cases of protozoal infections for demonstration of motile trophozoites.

Preservation of fecal specimens

In case of delay in examination, specimen may be placed in a suitable fixative. However, fecal specimen must not be incubated or frozen before a microscopic examination has been performed (Sapero and Lawless 1942, Isenberg 1995). Placing the specimen in a fixative helps to preserve the morphology of the parasites and inhibits the development of helminth eggs and/or larvae. A variety of preservatives are available which may be chosen based on the specific requirements. When a fixative is used, it is important that the fecal specimen is properly mixed with it. Some of the commonly used fixatives are formalin, merthiolate-iodine-formalin (MIF), sodium acetate-acetic acid-formalin (SAF), polyvinyl alcohol (PVA), Schaudinn's fluid, and the universal fixative. Most of the available fixatives allow for both routine stool examination processing and staining procedures. However, many of these fixatives may not be suitable when fecal immuno-assay is to be performed. In such situations, a fresh or frozen fecal specimen is preferred. Formalin, as a 10 percent (10%) aqueous solution, is one of the most commonly used preservative. However, it is helpful only for wet-mount examination which is considered to be significantly less accurate as compared to a permanent stained smear examination for demonstrating protozoal cysts. Aqueous solution of Formalin provides satisfactory results for preservation of the morphology of the parasites and for performing fecal immuno-assays for the detection of *Giardia lamblia* and *Cryptosporidium* spp. However, it is not suitable for preservation of morphology of trophozoites, and for performing permanent staining of fecal smears. Merthiolate-Iodine-Formalin (MIF) is another useful fixative that helps to preserve morphology of most of the parasites and their different stages in fecal specimens. Its ability to fix and stain the parasites simultaneously, gives an added advantage. However, the results obtained, in terms of preservation of morphology, are not as good as with other fixatives. Additionally, mercury, that is an environmental hazard, is an essential component of this fixative. It poses a significant problem for its safe disposal (Sapero and Lawless 1942, Isenberg 1995). Sodium acetate-Acetic acid-Formalin (SAF) is another fixative that can be used both for concentration and permanent stained smears. It is mercury-free and has a long shelf life. This preservative is particularly useful if fecal immunoassay is to be performed for the detection of *G. lamblia* and *Cryptosporidium* spp. However, it has poor adhesive properties and requires albumin-coated slides for smear preparation (Scholten and Yang 1974, Yang and Scholten 1977). Schaudinn's fluid is another useful preservative. It is generally considered as the gold standard for fixation and provides very good results on smears prepared with fresh fecal specimens, and morphology of protozoal cysts as well as trophozoites are very well preserved. However, it is not suitable for concentration procedures, and being a mercury-based solution, it poses an environmental hazard. Universal fixative (Total-fix) is another option which is available as a collection kit. This is a single-vial system for fecal specimens and helps to preserve the parasites. It offers very good preservation, and allows for demonstration of various parasites and their stages, including protozoa and helminths. Most of the procedures, including molecular tests, required for a parasitological work up, may be performed using this fixative. Since, it is free of mercury and formalin, it has added advantages.

Examination of fecal specimens

Macroscopic examination. A careful macroscopic examination of fecal specimens may provide important information which may help in reaching a diagnosis. Consistency of the specimen is one of the important features which should be noted during the initial examination. Based on the consistency, the specimen may be categorized as formed, soft or liquid. This information may indicate presence of certain parasites and their stages. In general, trophozoites are more likely to be seen in liquid, whereas both trophozoites and cysts may be observed in soft stool specimens. Presence of protozoal cysts is expected relatively more in formed fecal specimens. Therefore, a liquid specimen should be examined within 30 min, soft specimen may be examined within 1 hr, and formed specimen may be examined up to 24 hr after the passage of fecal specimen. However, these are not without exceptions; helminth eggs, coccidian oocysts or microsporidian spores, for example, may be demonstrated in any kind of specimen, although, the chances of detecting helminth eggs may be reduced in liquid specimens due to dilution. Macroscopic examination may also reveal adult worms or tapeworm segments, which can help in clinching the diagnosis. Color of the stool may also provide some useful information about the patient. For example, a dark colored specimen may be indicative of an upper gastro-intestinal bleeding, whereas appearance of bright red color may indicate lower gastro-intestinal bleeding. If both blood and mucus are observed, this may be indicative of some invasive parasitic infections. However, some medications and food items may also be responsible for abnormal coloration of the fecal specimens, as for example in patients on iron compounds, where specimen may show black discoloration.

Microscopic examination. This includes examination of direct wet-mounts, concentrated wet-mounts, and permanent stained smears (Clinical and Laboratory Standards Institute 2005, Melvin and Brooke 1982, Faust et al. 1938, Garcia et al. 1979, Garcia 1990, Levine and Estevez 1983, Markell and Voge 1981). In general, permanent stained smear examination allows for better appreciation of the parasite morphology and is preferred over a direct wet mount examination. Microscopic examination may demonstrate various parasites and their stages, including cysts, trophozoites, coccidian oocysts, spores of microsporidia, ova and larvae. Other significant findings could be red blood cells, pus cells, macrophages and charcot-leyden crystals which may provide some useful information/clues. It is needless to mention that microscopic examination of fecal specimens requires considerable experience and skill not only to identify and differentiate between different parasites, but also to distinguish them from the closely resembling artefacts in the specimen, such as pollen grains, plant cells, fungal spores, plant fibers, etc.

If direct wet-mount examination is to be performed, wet-mounts may be prepared by mixing small amount of fecal specimen (approximately two mg) with 0.85% saline on a slide and placing a 22 x 22 mm coverslip on it. If blood or mucus is observed on macroscopic examination of fecal specimen, a direct mount should be prepared and examined as early as possible. It is important to screen/examine entire coverslip under low power objective (10x). Any suspicious structure/materials identified in low magnification should immediately be examined under high power objective (40x). Even if no parasite is detected after low power screening, at least one-third of the cover slip area should be examined for parasites. The main/primary purpose of direct saline wet-mount examination is to demonstrate trophozoites, which may be detected and differentiated based on their characteristic motility. Due to the slow movement of the trophozoites, it is important that any suspicious structure encountered during examination is closely observed for at least fifteen seconds to detect any motility. Motility can also be induced by touching the edge of the slide with a hot coin or metallic substance. Besides making a saline wet mount, wet mount using iodine can be prepared by mixing fecal specimen with a weak iodine solution, such as Lugol's iodine. Iodine wet-mount helps in distinguishing internal structures of the parasitic trophozoites, cysts, oocysts and ova. Any protozoan parasite, detected during direct mount examination, needs to be confirmed and identified by examination of a permanent stained smear. Among the coccidian parasites, only the oocysts of *Cystoisospora belli* are generally identifiable easily. The oocysts of *Cryptosporidium* spp. and *Cyclospora cayetanensis* are usually too small to be identified during a direct wet mount examination, and require examination of smears stained with modified acid-fast staining method or by fecal immuno-assay. Ova or larvae of helminths, if observed, usually have characteristic morphological features by which they can be accurately identified. However, considering that both concentration procedures and permanent stained smear examination are superior to a direct wet-mount examination in demonstrating and identifying the parasites, it is not usually preferred in routine diagnosis of intestinal parasitic infections (Clinical and Laboratory Standards Institute 2005).

Concentration techniques. These methods allow for the detection of parasites even when the parasites are present in low numbers in the clinical specimens (Levine and Estevez 1983, Truant et al. 1981). The basic principle of these procedures is the use of centrifugation and differences in the specific gravity to separate parasites from the fecal debris (Clinical and Laboratory Standards Institute 2005, Garcia et al. 1979). The concentration techniques can be broadly classified into sedimentation techniques and floatation techniques. In general, sedimentation methods are more popular as they are simpler to perform and give good results. Formalin-ethyl acetate sedimentation concentration method is the most widely used concentration technique. Floatation techniques on the other hand, give a cleaner preparation; however, they may fail to concentrate some of the helminth eggs such as unfertilized eggs of *Ascaris lumbricoides* and eggs of trematodes.

Role of permanent staining. Any protozoa detected with direct wet mount or concentrated mount examination, should be confirmed by examining a permanent stained smear. This is important to correctly identify the observed protozoa, which may be difficult to distinguish from other pathogenic or non-pathogenic protozoan stages. Permanent staining allows for a better contrast to observe the characteristic morphological features. Additionally, the smears can be examined at higher magnification using oil immersion objective lens (100x), allowing for better appreciation of structural details. Several staining methods are available such as, trichrome staining and iron-hematoxylin staining. For routine diagnosis, trichrome staining or some modifications of iron-hematoxylin staining are preferred.

Quantitation of protozoa. Microscopic examination generally does not involve quantitation of protozoa. However, *Blastocystis hominis* is an exception where some indication of the load may be given by mentioning few, moderate or many. This is based on the number of parasites observed per ten fields at 100x magnification.

Demonstration of helminth eggs and larvae, coccidian oocysts or Microsporidian spores. This is based on examination of concentrated wet-mounts. Concentration wet mounts are also considered to be the best to demonstrate oocysts of *Cystoisospora belli.* Modified acid-fast staining or fecal immuno-assays are the recommended methods to demonstrate presence of *Cryptosporidium* spp. or *Cyclospora cayetanensis*. Modified acid-fast staining (Modified Kinyoun's staining) (Henriksen and Pohlenz 1981) is the most commonly used staining technique to demonstrate oocysts of coccidian parasites. Stained oocysts appear pink to red in color. They may even appear deep purple in color. This color variation may partly be due to differences in staining of fecal specimens due to variation in smear thickness. The oocysts of *Cryptosporidium* spp. and *Cyclospora cayetanensis* appear similar on examination. However, the size of *Cyclospora cayetanensis* is relatively larger (8 μm to 10 μm) as compared to *Cryptosporidium* spp. (4 μm to 6 μm). The former generally demonstrate a more variable staining and may have a 'wrinkled cellophane' like appearance. When the parasite load is low, the oocysts of *Cryptosporidium* spp. may be missed. Therefore, it is important to examine multiple specimens, preferably three specimens submitted on alternate days, to improve the sensitivity of detection of these parasites. There are some fecal immuno-assays available, which have a very good sensitivity for the detection of coccidian parasites and are particularly useful in diagnosing patients with low parasite load. Oocysts of coccidia may also be stained with Auramine 'O' stain. This is a low-cost, simple fluorescent staining method by which the stained oocysts appear as round fluorescent structures. This allows for a rapid screening of the fecal smears at low power objective (10x). This technique may also be used for screening un-concentrated direct specimens. Electron microscopy is considered to be the gold standard for demonstration of spores of microsporidia as well as for species determination. However, it requires highly skilled personnel, an expensive infrastructure, and needs more time. Therefore, it is not a preferred technique for routine diagnostic purpose. Modified trichrome staining is generally the most preferred method to demonstrate microsporidia in which microsporidian spores appear pink to red in color with a characteristic horizontal or diagonal stripe. However, due to their small size, they are often difficult to distinguish from some yeasts and bacteria which may also take up the stain.

Alternatives to microscopy. Although, microscopic examination of fecal specimens is the most popular and cost-effective method to demonstrate intestinal parasites, there are several other diagnostic modalities which may be used for diagnosing intestinal parasitosis. Many of these techniques are simple to perform and inexpensive, yet they are not routinely used (Garcia 2009, Isenberg 2004, Isenberg 1995, Garcia 2010). These include fecal cultures for protozoa and helminths. Protozoa, such as Entamoeba histolytica and Giardia lamblia can be cultivated using various special media. For helminths, cultivation basically refers to various techniques which may be used to concentrate the larval stages of nematodes. This is most commonly used to demonstrate larvae of *Strongyloides stercoralis*. Fecal cultures are particularly useful in cases with low parasite load or when we need to differentiate larvae of *Strongyloides stercoralis*, hookworm or *Trichostrongylus* spp., or for specific identification of nematodes when their eggs are indistinguishable on microscopic examination. Some of the commonly used fecal culture methods are (1) Harada-Mori method, (2) Slant culture method, (3) Charcoal culture, (4) Baermann technique, and (5) Agar plate culture. It is important not to refrigerate or add any fixative to the fecal specimens when fecal cultures are being considered (Garcia 2009, Harada and Mori 1955, Little 1966, Watson and Al-Hafidh 1957). Agar plate culture method is considered to be the most sensitive, and is recommended for recovery of *Strongyloides stercoralis* larvae. This is because up to half of the infected patients may fail to demonstrate the larvae in fecal specimens. Considering the high risk of complications associated with steroid treatment in patients infected with *Strongyloides stercoralis*, it is imperative that this infection has been ruled out with high level of certainty to avoid improper management of these cases (Arakaki et al. 1990, Jongwutiwes et al. 1999, Enes Ede et al. 2011).

Quantitation of helminths. Methods are available by which worm burden may be estimated. However, a correlation between egg production and number of adult worms present may be relevant only in cases of infections with *Ascaris lumbricoides, Trichuris trichiura,* and Hookworms. This information may be useful to assess the severity of infection or to evaluate the response to treatment. Some of the methods which may be used for this purpose are (1) Direct smear method of Beaver, (2) Dilution egg count (Stoll count), (3) Kato-Katz thick smear, and (4) McMaster method (Melvin and Brooke 1982, Beaver 1949, Stoll and Hausheer 1926, Levecke et al. 2011). However, with the availability of effective treatment for most of the intestinal helminthic infections, these egg-counting techniques are not considered to be of much clinical significance.

Other specimens for intestinal parasitic infections

Besides feces, other specimens may also be useful for diagnosing intestinal parasitic infections. For example, eggs of *Enterobius vermicularis* are usually not detected in feces as the female worm deposits the eggs in perianal region; therefore, in suspected cases of Enterobiasis, a perianal sampling is preferred. This may be done using cellophane tape preparation (Garcia 2009, Brooke and Melvin 1969, Graham 1941) or anal swabs (Melvin and Brooke 1974). Best results are achieved with morning specimens (collected before the patient takes a bath or uses the washroom), and when at least 4–6 consecutive specimens are screened.

Sigmoidoscopy-material is another specimen which may be used to diagnose intestinal amebiasis, giardiasis, or cryptosporidiosis, when detailed fecal examination has failed to demonstrate the parasites on multiple occasions. It is important to note that the specimen obtained by sigmoidoscopy needs to be aspirated or scraped from the mucosal surface, and at least six representative areas should be sampled (Garcia 2009, 2010). The specimen should be processed immediately and direct wet mounts and permanent stained smears should be examined. Depending on the investigations planned on the specimen, different strategies should be adopted for processing of the specimen as is done with fecal specimens. However, due to the invasive nature of this sampling method, fecal immuno-assays are considered to be very good alternatives to sigmoidoscopy based methods.

Similarly, duodenal drainage/aspiration may be performed to diagnose cases of Giardiasis, Strongyloidiasis, or Cryptosporidiosis. The specimen should be examined within 2 hr after it is collected, or it should be preserved in suitable fixatives. Since trophozoites of *Giardia lamblia* are commonly trapped in mucus, it is important to centrifuge the specimen and examine the sediment. A simpler option to obtain specimen from duodenum is 'The duodenal capsule technique' (Beal et al. 1970), also known as 'The Entero test'. This method makes use of a gelatin capsule with a length of nylon string coiled inside. One end of the string is free and protrudes out of the capsule. This free end is taped to the side of the face of the patient and the patient is instructed to swallow the capsule. Eventually the capsule dissolves and the string which reaches the duodenum gets coated with the duodenal contents including mucus. After a period of 4 hr, the string is withdrawn and the material sticking at the end is scraped off in to a petri dish. Presence of yellow-green material on the string, or a low pH of the specimen helps to ensure that the string had reached duodenum and the specimen is representative of duodenal contents. Specimen is then immediately examined to look for any motile trophozoites. In case a delay in examination is expected, a suitable preservative should be added followed by a permanent staining procedure.

Parasitic Infections of The Uro-Genital Tract

In patients with suspected parasitic infections of uro-genital tract, different approaches may be required depending upon the target parasite. For suspected trichomoniasis cases, wet mount examination of vaginal and urethral discharges, prostatic secretions or urinary sediment are useful. Concentration of the urine specimen by centrifugation, improves the sensitivity of microscopic examination as does the examination of multiple vaginal secretion specimens. For the diagnosis of *Trichomonas vaginalis* infections, molecular techniques are now becoming popular due to their high sensitivity and specificity (Heine et al. 1997, Paterson et al. 1998, Ginocchio et al. 2012). Presence of helminthic infection of urinary tract may be confirmed by microscopic examination of urine specimens. Centrifugation of the specimen, or better still, the use of membrane filtration techniques improve the sensitivity of detecting these parasites, including microfilariae and *Schistosoma hematobium* eggs (Feldmeier et al. 1979).

Parasitic Infections of The Respiratory Tract

For respiratory tract infections, such as pneumonia, pneumonitis, and Loeffler's syndrome, expectorated sputum may be used. Sputum specimen may be examined as a wet-mount preparation to demonstrate larvae (e.g., *Ascaris lumbricoides, Strongyloides stercoralis*), eggs (e.g., *Paragonimus westermanii*), hooklets and scolices (e.g., *Echinococcus granulosus*), oocysts (e.g., *Cryptosporidium* spp.), trophozoites (e.g., *Entamoeba histolytica*), microsporidian spores. In cases with suspected *Cryptosporidium* infection of the respiratory tract, modified acid-fast staining or immunoassays may be helpful (Current and Garcia 1991, Garcia 1999, Garcia and Shimizu 1997, Siddons et al. 1992).

To ensure reliable laboratory results, it is important that sputum specimen, should be collected properly. The patient should be instructed to collect sputum specimen, preferably an early morning sample, after deep coughing, so that the specimen collected is of good quality with minimal mixing with saliva. Patient instructions should include proper mouth washing before the specimen is collected. After the specimen is collected, it should be delivered to the laboratory as soon as possible and then complete examination should be performed, preferably using viscous or any blood tinged part of the specimen.

Parasitic Infections of Blood

Besides intestinal infections, infections due to haemo-parasites, is the other common and important group of parasitic infections. A number of parasites have at least some of the stages in their life-cycle passing through blood. These stages may be demonstrated by examination of blood specimens, either using whole blood or after concentration procedures. Some of the important blood parasites include *Plasmodium* spp., *Trypanosoma* spp., *Babesia* spp., *Leishmania donovanii,* and microfilariae of some filarial nematodes. For most of these parasites, examination of stained thin and thick peripheral blood films is considered to be the most useful and recommended diagnostic technique. The blood films may be prepared directly from the specimens, specimens with anti-coagulant, or from concentrated blood specimens. The most preferred stains used are Romanowsky's stains and include Giemsa, Wright's, Field's, and Leishman's stain.

Thick and thin blood films

It is a common and recommended approach that both thin and thick smears are prepared from each blood specimen received for parasitological work up. A larger amount of blood is used in preparation of a thick film, and this increases the sensitivity of the microscopic examination, by increasing the possibility of detecting parasites in case of low parasitemia (Field et al. 1963). Thin smear, on the other hand, is more useful for species identification, particularly in cases of malaria, where it allows for better appreciation of morphological details of the parasites as well as infected red blood cells. Whenever a patient is suspected to have a parasitic infection with blood stream involvement, blood specimen should be immediately withdrawn and sent for examination. Since parasite may not be detected by a single peripheral blood screening, at least three specimens at different points of time should be examined on different days, before a patient is declared to be free of infections due to blood parasites. Usually fresh blood with anticoagulant (preferably EDTA) is used. However, if finger prick method is used for obtaining a blood specimen for examination, it should be ensured that it is free flowing, and there is minimal dilution with tissue fluids to affect the concentration of parasites in the specimen. As soon as a blood specimen is received for parasitological investigation, it should be immediately used to prepare blood films, and fixed with methyl alcohol followed by appropriate staining and examination. Delay in processing of the specimens may affect the morphology or numbers of the parasites and may even lead to washing off of the films during the staining process (Isenberg 1995, 2004, Garcia 2010, Clinical and Laboratory Standards Institute 2000, Kokoskin 2001). Once the thin smears are stained, they should be screened at low power objective. This initial screening may reveal some parasites such as microfilariae, which may even be an accidental finding in some cases, with different etiology responsible for the patient's complaints. It is important that entire smear is screened so as not to miss any finding. To demonstrate the presence of malarial parasites, the tail end that is the end of the thin smear should be examined. The red blood cells, at this end are present as a single distinctive layer of cells, and it is relatively easier to detect and identify the parasite and its species. Before a thin film is reported as negative, at least 200 to 300 fields should be examined with oil immersion objective lens at 1000x magnification. In thick films, parasites may be detected even if there are only 5–20 parasites per µL of specimen. A routine 100 field examination protocol may miss up to 20% of infections. Therefore, at least 300 fields should be examined, and for each specimen both thin and thick films should be screened. If one set of peripheral blood films is negative, it does not necessarily rule out malaria, and examination of at least three sets collected on different days should be done (Hänscheid 1999). For stained thick smears, again it is important to screen the entire smear at low power magnification to rule-out microfilariae. Examination for malarial parasites or Trypanosomes is performed at 1000x magnification using oil immersion objective lens. At least 100 fields should be examined before reporting a smear as negative.

Parasite count

When a specimen is positive for the presence of malarial parasite, a parasite count is performed to assess the parasite load and to correlate with severity of infection. This helps in monitoring the response to specific treatment by comparing the counts over a period at different intervals, generally at 24, 48, and 72 hr after the treatment has been started. This may help in identifying the cases with drug-resistant strains and guiding further patient management. The parasite count may be performed using different recommended methods. One of the method uses red blood cell (RBC) count as the reference, and the parasite count is reported as the percentage of infected RBCs per 100 RBCs counted (Clinical and Laboratory Standards Institute 2000). This method is applicable when thin smear is used for examination. Another method uses normal white blood cells (WBCs) as the reference. Here, the number of infected RBCs in the smear per 100 WBCs is counted. The count can then be used to derive the parasite count per microliter of blood using the total leukocyte count of the patient. When this method is used, blood specimen for both the peripheral smear examination as well as cell counts, must be submitted simultaneously. After the parasite counts are estimated, it is important to check whether they are indicative of serious malaria, which is represented by 2–5% parasitemia (1,00,000 to 2,50,000 parasites per µL of blood). These cases need special attention as they are at high risk of complications and mortality. When parasitemia is greater than or equal to 10% (5,00,000 parasites per µL), there is a high risk of mortality and exchange transfusion should be considered (Hänscheid 1999).

Alternatives to thick and thin smear examination

The microscopic examination of stained thin and thick blood films remains the mainstay for the diagnosis of malaria, and is considered as the gold standard for this purpose. The main drawbacks of thin and thick smears include the requirement of expertise to identify and differentiate the malarial parasites and it is time consuming. There are several other alternatives available, which may be used for malaria diagnosis. These include Rapid Diagnosis Tests (RDTs) to demonstrate malaria antigen or antibodies besides Quantitative Buffy Coat (QBC) assay, PCR assays, and automated blood cell analyzers.

Rapid Diagnosis Test (RDT) kits

RDTs are an important group of assays which are commonly used. These are available both for malaria antigen and antibody detection. The antigen based assays are based on the detection of either *Plasmodium* LDH (Lactate De-Hydrogenase) (pLDH), *Plasmodium falciparum* HRP 2 (Histidine-Rich Protein 2) (pf HRP 2), or *Plasmodium* aldolase. Aldolase based diagnosis generally has a low sensitivity. Most of these assays can give the report within 15 minutes and are simple to perform and read. The commercial kits are available in different combinations, such as pf HRP 2 and pan LDH (detects LDH isotopes for all the human *Plasmodium* species), pf LDH and pan LDH, pf HRP2 and *Plasmodium vivax* LDH, and many more. pf HRP2 is produced by asexual stages and gametocytes of *P. falciparum* and is expressed on the surface of red cell membrane. It may be detectable in the blood for at least 28 days after the patient is started on anti-malarial therapy. Also, due to strain variations, there might be some differences in the pf HRP2 sequences in different countries and this may lead to variation in the results obtained with HRP 2 based assays (Baker et al. 2005). The other drawback of the HRP 2 based assays is related to its persistence in blood for a long time even after treatment initiation. This may be responsible for false positive results reported in many patients even after treatment. pLDH is a soluble glycolytic enzyme produced by the sexual and asexual stages of the live parasites. It is present inside the infected red cells and is also released from the infected cells. The advantage of pLDH is that it is detectable in the circulation for a shorter time, up to 10 days, after initiation of the treatment. Hence, there is lesser chance of false positive results in patients who have received anti-malarial treatment. Different isomers are available for each of five human *Plasmodium* species and depending on the need they may be used. However, pan LDH based assays perform well mainly against *P. vivax*, and may not provide accurate results with other *Plasmodium* spp. (Makler et al. 1998, Palmer et al. 1998). The test gives positive results only when viable parasites are present. The sensitivity and specificity is similar to HRP 2 based assays with a sensitivity of approximately 95% at parasitemia of more than 100 per microliter of blood but just around 50% for counts less than 50 per microliter of blood. Therefore the use of RDTs may improve the sensitivity of diagnosing malaria in areas with low density parasitaemia, and may be beneficial in places where experts in microscopy based malaria diagnosis are not available. However, microscopy is still essential for species identification, identification of mixed infections, confirmation of the diagnosis and to estimate parasite count.

Quantitative Buffy Coat (QBC) assay

QBC assay is another alternative for malaria diagnosis. This involves microscopic examination of 50 to 60 μL of blood specimen after a process of micro-hematocrit centrifugation in QBC malaria tubes. These tubes are coated with a fluorescent stain (acridine orange) and certain anticoagulants. The method further involves high speed centrifugation of the blood specimen in QBC tubes. This causes density-based differential separation of different blood cells, and parasitized and normal cells, which become concentrated in different layers. The fluorescence imparted to the parasites by acridine orange dye, further helps to increase the sensitivity of the test. The sensitivity of the assay has been reported to be in a wide range of 55% to more than 90% compared to thin and thick blood film microscopy. The major drawbacks associated with this method are (1) high cost of disposables and equipment, (2) difficulty in species identification, (3) difficulty in identifying mixed infections, and (4) difficulty in determination of parasite load (Hänscheid 1999). This method may also be used to demonstrate other blood stream infections, such as Filariasis and tick-borne relapsing fever, with high level of sensitivity (Wang 1998, Van Dam et al. 1999).

Molecular tests

Many PCR based assays are available which may be used to accurately identify the *Plasmodium* spp. especially when there is doubt. These assays have higher sensitivity than microscopy or RDTs and are particularly useful in low parasitemia malaria cases. Also, they may be extremely helpful in diagnosing and differentiating cases of mixed infections due to multiple species of *Plasmodium*. In such cases, microscopy or RDTs may fail to provide the answers. The main problems with adoption of molecular techniques in routine malaria diagnosis include (1) requirement of significantly longer time for the assay, (2) need for technical expertise, and (3) higher costs. At present, the molecular techniques are performed mainly in referral centers or under special circumstances.

Diagnosis of leishmaniasis

Leishmaniasis is an important public health problem in several countries. Diagnosis of leishmaniasis has traditionally depended upon microscopy based methods. In cases of visceral leishmaniasis, specimens such as bone marrow aspirate and splenic aspirate are used, while for cutaneous leishmaniasis, skin biopsy specimen is used. Smears are prepared and examined for the presence of *Leishmania donovani* bodies (LD bodies) after staining with Romanowsky's stain (Giemsa, Leishman's, etc.). However, many alternatives are now available which have helped in improving the sensitivity of diagnosis and can be performed without dependency on any invasive procedure. The rK-39 immuno-chromatographic test (rK-39 ICT) is a highly sensitive and specific assay (both around 100%) for diagnosing visceral leishmaniasis and post-kala azar dermal leishmaniasis (PKDL). It is based on the demonstration of anti-*leishmania* antibodies in patient's serum, and is a rapid and reliable method for diagnosis of leishmaniasis (Mathur et al. 2005). Cultivation of the parasite and molecular techniques are other available options.

Diagnosis of lymphatic filariasis

Diagnosis of lymphatic filariasis involves examination of stained thick and thin blood films. Several concentration methods are also used, which include centrifugation of blood specimen with a suitable lysing agent (e.g., saponin) to lyse red blood cells and membrane filtration methods. Wet-mounts prepared from freshly collected anti-coagulated blood or after concentration of blood specimen may be microscopically examined. Use of QBC assay can also demonstrate the microfilariae. In some cases, sensitivity of microscopic demonstration of microfilariae may be improved by collecting the blood specimen between 10 pm and 2 am. This is particularly useful in areas where microfilariae may show nocturnal periodicity. Other option in such cases is to use Hetrazan or Diethylcarbamazine (DEC) provocative test. This test involves oral administration of 100 mg of DEC to an adult patient which induces the appearance of microfilariae in the peripheral blood. Blood specimen is collected after waiting for 30–45 min and is examined after concentration. Some RDT kits for detection of antigens of *Wuchereria bancroftii* are also available and have high sensitivity and specificity. Depending on the clinical presentation, microfilariae may sometimes be demonstrated in other specimens such as lymph node biopsy and urine in patients with chyluria.

Parasitic Infections of Other Organ-Systems

Parasites may cause infection in various organ-systems of human body. Aspirates from different tissues and organs may be used for diagnosing parasitic diseases. Common aspirates which are submitted for parasitological work up include fine-needle aspirates, broncho-alveolar lavage fluid, tracheal aspirate, and bronchial washings. The material may be used for examination by preparing wet-mounts or using appropriate staining methods. It may also be subjected to culture for isolating specific parasites in certain cases. The processing and staining methods used generally depend up on the organ system involved and the parasites suspected. For example, liver aspirates may be useful in demonstrating the amebic trophozoites in cases of amebic liver abscess (ALA), or hooklets and scolices in cases of hydatid disease. It is important that material is collected from the margin of the abscess in case of suspected amebic liver abscess, to improve the probability of recovering the parasites. However, in many cases of extra-intestinal amebiasis, it may be difficult to demonstrate the trophozoites and thus serological tests are recommended for the diagnosis. When a patient is suspected to have hydatid disease, examination of aspirated cyst fluid may be helpful to confirm the diagnosis. However, aspiration of a hydatid cyst has associated risks which may have serious and fatal consequences. Therefore, this procedure must be performed very carefully and only during open surgical procedures where cyst removal is being done. The aspirate can be examined as wet-mounts directly from the aspirate or from the sediment obtained after centrifugation. However, the hooklets and scolices may not be demonstrable in all the cases, as sometimes, the cysts are sterile with no brood capsules. In such cases, a histopathological examination of the cyst wall may be used for confirming the diagnosis. In suspected cases of primary amebic encephalitis (PAM) or granulomatous amebic encephalitis (GAE), caused by Free-living amebae (FLA), a direct wet mount examination of un-centrifuged spinal fluid is performed. This may demonstrate motile trophozoites of free living amebae. Smears may be prepared from the specimen and may be examined after staining with stains such as, Giemsa and Wright's, to detect and identify these parasites. In case of ocular infections, calcofluor white staining may be used to demonstrate cysts or trophozoites of *Acanthamoeba* spp. or spores of microsporidia. Many patients may present with parasitic infections of the skin. Examination of appropriately stained impression smears from the lesions or biopsy specimens may help to reveal the parasitic cause. In case of ulcers, specimen should be collected from below the ulcer bed, through the uninvolved skin. Also, the site to be sampled should be properly cleaned to avoid contamination with commensal organisms. This is especially important when specimen may be subjected to culture methods. Biopsy specimens may be examined after preparing impression smears, using different histopathological procedures, and/or by using various staining techniques depending on the parasite suspected in the lesions. Accurate diagnosis is largely dependent upon the proper sampling, site selection and presence of the tissue in sufficient quantities. Examination of multiple tissue specimens is often necessary to detect tissue parasites. The specimens may also be used for *in vitro* culture methods when indicated.

Alternative Methods for Diagnosis of Parasitic Infections

Molecular techniques

Molecular techniques, such as PCR, Real-time PCR, and Multiplex PCR assays have a very high sensitivity in diagnosing many parasitic infections. These may be used when the routine investigations fail to establish a definitive diagnosis. In addition, there are several molecular typing techniques available, which may be used for epidemiological purposes. However, due to the requirement of high level technical expertise, higher cost and increased turn-around time, their application in routine diagnosis is limited.

Serological methods

In many parasitic infections, direct demonstration of parasites may not be possible. Several serological methods are available with high sensitivity and specificity, and are used for routine diagnosis of parasitic infections. Examples include ELISA kits, complement fixation tests (CFT), immuno-chromatography tests (ICT), and Immuno-fluorescent Assay (IFA) tests for demonstration of specific IgM or IgG antibodies against *Toxoplasma gondii*, *Toxocara canis*, *Echinococcus granulosus*, *Entamoeba histolytica*, *Leishmania donovani*, and malaria parasites. Being dependent upon the detection of antibodies, these are faced with disadvantages which are inherent to many other antibody based assays. The major drawbacks include (1) antibodies may not be demonstrable during the initial few days after the infection, (2) in endemic areas, presence of antibodies may be due to previous exposures and may not represent an active infection, (3) antibodies may persist for long time even after successful treatment and may not have any prognostic value, and (4) in patients who are immune-suppressed, antibody response may not be high enough to be detected by these assays.

Culture of parasites and xenodiagnosis

Besides the common methods used for parasitological diagnosis, there are certain specialized methods which are primarily used for research purposes; however, these may sometimes be used for diagnosis in difficult cases. These methods generally involve recovery of parasites and are not usually feasible in routine diagnostic laboratories.

Culture of parasites

The culture methods are available for some parasites such as *Entamoeba histolytica*, *Giardia lamblia*, *Acanthamoeba* spp., *Naegleria fowleri*, *Trichomonas vaginalis*, *Toxoplasma gondii*, *Leishmania* spp., and *Blastocystis hominis*. The culture techniques may involve growing the parasite in association with unknown microbiota (xenic culture), or with a single known bacterium (mono-xenic culture), e.g., free-living amoebae, or using artificial media with no bacteria (axenic culture). Tissue culture techniques may also be used for cultivating some parasites such as, free-living amoebae, *Cryptosporidium* spp., and Microsporidia. However, for majority of these infections, the role of cultures is limited to research (Isenberg 2004, Garcia 2010, Arrowood 2002, Clark and Diamond 2002, Schuster 2002a,b, Schuster and Sullivan 2002, Taylor and Baker 1968, Visvesvara 2002, Visvesvara and Garcia 2002).

Diagnosis of *Trichomonas vaginalis*, which is responsible for sexually transmitted infections (STI), involves collection of vaginal exudates in case of females, and semen, urethral specimens or urine from males. In cases where urine is used, first voided morning urine specimen is preferred. Culture is considered to be the most sensitive method for diagnosis; however, it is labor-intensive and time-consuming. Therefore, molecular techniques are now preferred.

Cultures may also be used to diagnose infections caused by hemoflagellates. Examples include, Trypanosomes and *Leishmania* spp., where a common approach is to use Novy MacNeal-Nicolle (NNN) medium for cultivating the parasites. In patients with toxoplasmosis, some patients may have a low immune response. These patients may be difficult to diagnose with routine serological investigations. Histo-pathological examinations have their own limitations in terms of low sensitivity. In such patients, tissue culture methods may be of significant help. Culture methods have also been developed for *Plasmodium* spp. However, *P. falciparum* and *P. knowlesi* are the only human species, where all the parasite stages have been successfully cultured *in vitro*. Since many rapid, as well as, sensitive and specific diagnostic tests are available for malaria diagnosis, cultures are not usually indicated. Their main application remains in research on these important human parasites (Schuster 2002b).

Animal inoculation

With good diagnostic modalities available for most of the common human parasitic infections, use of animals in diagnostic parasitology has become very limited. Animal inoculation techniques are now not used for routine parasitological workup by most of the clinical laboratories.

Xenodiagnosis

Xenodiagnosis is another method for recovery of parasites which involves use of an arthropod host. For example, reduviid bugs may be used in the diagnosis of Chagas' disease. Here, the reduviid bugs are allowed to feed on the blood of a patient with suspicion of this parasitic infection. After 1–2 months, the arthropod's feces is examined for the developmental stages of *Trypanosoma cruzi* (Maekelt 1964). This process is continued for a period of 3 months before reporting the specimen as negative. Xenodiagnosis is very rarely used for diagnostic purposes.

References

Arakaki, T., M. Iwanaga, F. Kinjo, A. Saito, R. Asato and T. Ikeshiro. 1990. Efficacy of agar-plate culture in detection of *Strongyloides stercoralis* infection. J. Parasitol. 76: 425–428.

Arrowood, M.J. 2002. *In vitro* cultivation of *Cryptosporidium* species. Clin. Microbiol. Rev. 15: 390–400.

Baker, J., J. McCarthy, M. Gatton, D.E. Kyle, V. Belizario, J. Luchavez et al. 2005. Genetic diversity of *Plasmodium falciparum* Histidine-Rich Protein 2 (PfHRP2) and its effect on the performance of PfHRP2-based rapid diagnostic tests. J. Infect. Dis. 192: 870–877.

Beal, C.B., P. Viens, R.G.L. Grant and J.M. Hughes. 1970. A new technique for sampling duodenal contents: demonstration of upper small-bowel pathogens. Am. J. Trop. Med. Hyg. 19: 349–352.

Beaver, P.C. 1949. A nephelometric method of calibrating the photoelectric meter for making egg counts by direct fecal smear. J. Parasitol. 35: 13.

Brooke, M.M. and D. Melvin. 1969. Morphology of diagnostic stages of intestinal parasites of man. U.S. Department of Health, Education, and Welfare publication (HSM) 72-8116. US Government Printing Office, Washington, DC.

Cartwright, C.P. 1999. Utility of multiple stool specimen ova and parasite examinations in a high-prevalence setting. J. Clin. Microbiol. 37: 2408–2411.

Clark, C.G. and L.S. Diamond. 2002. Methods for cultivation of luminal parasitic protists of clinical importance. Clin. Microbiol. Rev. 15: 329–341.

Clinical and Laboratory Standards Institute. 2000. Laboratory diagnosis of blood-borne parasitic diseases. Approved guideline M15-A. Clinical and Laboratory Standards Institute, Villanova, PA.

Clinical and Laboratory Standards Institute. 2005. Procedures for the recovery and identification of parasites from the intestinal tract. Approved guidelines M28-2A. Clinical and Laboratory Standards Institute, Villanova, PA.

College of American Pathologists. 2012. Commission on Laboratory Accreditation Inspection Checklist. College of American Pathologists, Chicago, IL.

Committee on Education, American Society of Parasitologists. 1977. Procedures suggested for use in examination of clinical specimens for parasitic infection. J. Parasitol. 63: 959–960.

Current, W.L. and L.S. Garcia. 1991. Cryptosporidiosis. Clin. Microbiol. Rev. 4: 325–358.

Enes Ede, J., J.N. Souza, R.C. Santos, E.S. Souza, F.L. Santos, M.L. Silva et al. 2011. Efficacy of parasitological methods for the diagnosis of *Strongyloides stercoralis* and hookworm in faecal specimens. Acta Trop. 120: 206–210.

Faust, E.C., J.S. D'Antoni, V. Odom, M.F. Miller, C. Peres, W. Sawitz et al. 1938. A critical study of clinical laboratory techniques for the diagnosis of protozoan cysts and helminth eggs in feces. Am. J. Trop. Med. 18: 169–183.

Feldmeier, H., U. Bienzle, M. Dietrich and H.J. Sievertsen. 1979. Combination of a viability test and a quantification method for schistosoma hematobium eggs (filtration-trypan blue staining-technique). Tropenmed. Parasitol. 30: 417–422.

Field, J.W., A.A. Sandosham and Y.L. Fong. 1963. The microscopical diagnosis of human malaria 1. A morphological study of the erythrocytic parasites in thick blood films. Institute for Medical Research, Kuala Lumpur, Malaya.

Garcia, L., T. Brewer and D. Bruckner. 1979. A comparison of the formalin ether concentration and trichrome stained smear methods for the recovery and identification of intestinal protozoa. Am. J. Med. Technol. 45: 932–935.

Garcia, L.S. and M. Voge. 1980. Diagnostic clinical parasitology. I. Proper specimen collection and processing. Am. J. Med. Technol. 46: 459–467.

Garcia, L.S. 1990. Laboratory methods for diagnosis of parasitic infections. pp. 776–861. *In*: Baron, E.J. and S.M. Finegold (eds.). Bailey & Scott's Diagnostic Microbiology, 8th ed. The C.V. Mosby Co., St. Louis, MO.

Garcia, L.S. and R.Y. Shimizu. 1997. Evaluation of nine immunoassay kits (enzyme immunoassay and direct fluorescence) for detection of *Giardia lamblia* and *Cryptosporidium parvum* in human fecal specimens. J. Clin. Microbiol. 35: 1526–1529.

Garcia, L.S. 1999. Practical Guide to Diagnostic Medical Parasitology. ASM Press, Washington DC.

Garcia, L.S., J.W. Smith and T.R. Fritsche. 2003. Cumitech 30A, Selection and Use of Laboratory Procedures for Diagnosis of Parasitic Infections of the Gastrointestinal Tract. Coordinating ed., L.S. Garcia. ASM Press, Washington, DC.

Garcia, L.S., S.P. Johnston, A.J. Linscott and R.Y. Shimizu. 2008. Cumitech 46, Laboratory Procedures for Diagnosis of Blood-borne Parasitic Diseases. Coordinating ed., L.S. Garcia. ASM Press, Washington, DC.

Garcia, L.S. 2009. Practical Guide to Diagnostic Medical Parasitology, 2nd ed. ASM Press, Washington, DC.

Garcia, L.S. (ed.). 2010. Clinical Microbiology Procedures Handbook, 3rd ed. ASM Press, Washington, DC.

Garcia, L.S. 2016. Diagnostic Medical Parasitology, 6th ed. ASM Press, Washington, DC.

Ginocchio, C.C., K. Chapin, J.S. Smith, J. Aslanzadeh, J. Snook, C.S. Hill et al. 2012. Prevalence of *Trichomonas vaginalis* and Coinfection with *Chlamydia trachomatis* and *Neisseria gonorrhoeae* in the United States as Determined by the Aptima *Trichomonas vaginalis* nucleic acid amplification assay. J. Clin. Microbiol. 50: 2601–2608.

Graham, C.F. 1941. A device for the diagnosis of *Enterobius* infection. Am. J. Trop. Med. 21: 159–161.

Hänscheid, T. 1999. Diagnosis of malaria: a review of alternatives to conventional microscopy. Clin. Lab. Haematol. 21: 235–245.

Harada, U. and O. Mori. 1955. A new method for culturing hookworm. Yonago Acta Med. 1: 177–179.

Heine, R.P., H.C. Wiensfeld, R.L. Sweet and S.S. Witkin. 1997. Polymerase chain reaction analysis of distal vaginal specimens: a less invasive strategy for detection of *Trichomonas vaginalis*. Clin. Infect. Dis. 24: 985–987.

Henriksen, S.A. and J.F.L. Pohlenz. 1981. Staining cryptosporidia by a modified Ziehl-Neelsen technique. Acta Vet. Scand. 22: 594–596.

Isenberg, H.D. (ed.). 1995. Essential procedures for clinical microbiology. American Society for Microbiology, Washington, DC.

Isenberg, H.D. (ed.). 2004. Clinical Microbiology Procedures Handbook, 2nd ed. ASM Press, Washington, DC.

Jones, J.L., A. Lopez, S.P. Washquist, J. Nadle and M. Wilson, The Emerging Infections Program FoodNet Working Group. 2004. Survey of clinical laboratory practices for parasitic diseases. Clin. Infect. Dis. 38: S198–S202.

Jongwutiwes, S., M. Charoenkorn, P. Sitthichareonchai, P. Akaraborvorn and C. Putaporntip. 1999. Increased sensitivity of routine laboratory detection of *Strongyloides stercoralis* and hookworm by agar-plate culture. Trans. R. Soc. Trop. Med. Hyg. 93: 398–400.

Kokoskin, E. 2001. The Malaria Manual. McGill University for Tropical Diseases, Montreal, Canada.

Levecke, B., J.M. Behnke, S.S.R. Ajjampur, M. Albonico, S.M. Ame, J. Charlier et al. 2011. A comparison of the sensitivity and fecal egg counts of the McMaster egg counting and Kato-Katz thick smear methods for soil-transmitted helminths. PLoS Negl. Trop. Dis. 5: e1201. Doi: 10.1371/journal.pntd.0001201.

Levine, J.A. and E.G. Estevez. 1983. Method for concentration of parasites from small amounts of feces. J. Clin. Microbiol. 18: 786–788.

Little, M.D. 1966. Comparative morphology of six species of *Strongyloides* (Nematoda) and redefinition of the genus. J. Parasitol. 52: 69–84.

Maekelt, G. 1964. A modified procedure of xenodiagnoses of Chagas' disease. Am. J. Trop. Med. Hyg. 13: 11–15.

Makler, M.T., R.C. Piper and W.K. Milhous. 1998. Lactate dehydrogenase and the diagnosis of malaria. Parasitol. Today 14: 376–377.

Markell, E.K. and M. Voge. 1981. Medical Parasitology, 5th ed. The W.B. Saunder's Co., Philadelphia, PA.

Mathur, P., J. Samantaray and N.K. Chauhan. 2005. Evaluation of a rapid immunochromatographic test for diagnosis of kala-azar and post kala-azar dermal leishmaniasis at a tertiary care centre of north India. Indian J. Med. Res. 122: 485–490.

McVicar, J.W. and J. Suen. 1994. Packaging and shipping biological materials. pp. 239–246. *In*: Fleming, D.O., J.H. Richardson, J.J. Tulis and D. Vesley (eds.). Laboratory Safety: Principles and Practices, 2nd ed. ASM Press, Washington, DC.

Melvin, D.M. and M.M. Brooke. 1974. Laboratory procedures for the diagnosis of intestinal parasites. US Government Printing Office, Washington, DC.

Melvin, D.M. and M.M. Brooke. 1982. Laboratory Procedures for the Diagnosis of Intestinal Parasites, 3rd ed. U.S. Department of Health, Education, and Welfare publication (CDC) 82-8282. Government Printing Office, Washington, DC.

Palmer, C.J., J.F. Lindo, W.I. Klaskala, J.A. Quesada, R. Kaminsky, M.K. Baum et al. 1998. Evaluation of the OptiMAL test for rapid diagnosis of *Plasmodium vivax* and *Plasmodium falciparum* malaria. J. Clin. Microbiol. 36: 203–206.

Paterson, B.A., S.N. Tabrizi, S.M. Garland, C.K. Fairley and F.J. Bowden. 1998. The tampon test for trichomoniasis: a comparison between conventional methods and a polymerase chain reaction for *Trichomonas vaginalis* in women. Sexually Transmitted Infections 74: 136–139.

Sapero, J.J. and D.K. Lawless. 1942. The MIF stain preservation technique for the identification of intestinal protozoa. Am. J. Trop. Med. Hyg. 2: 613–619.

Sawitz, W.G. and E.C. Faust. 1942. The probability of detecting intestinal protozoa by successive stool examinations. Am. J. Trop. Med. 22: 131–136.

Scholten, T.H. and J. Yang. 1974. Evaluation of unpreserved and preserved stools for the detection and identification of intestinal parasites. Am. J. Clin. Pathol. 62: 563–567.

Schuster, F.L. 2002a. Cultivation of pathogenic and opportunistic free living amebas. Clin. Microbiol. Rev. 15: 342–354.

Schuster, F.L. 2002b. Cultivation of *Plasmodium* spp. Clin. Microbiol. Rev. 15: 355–364.

Schuster, F.L. and J.J. Sullivan. 2002. Cultivation of clinically significant hemoflagellates. Clin. Microbiol. Rev. 15: 374–389.

Siddons, C.A., P.A. Chapman and B.A. Rush. 1992. Evaluation of an enzyme immunoassay kit for detecting cryptosporidium in faeces and environmental samples. J. Clin. Pathol. 45: 479–482.

Stoll, N.R. and W.C. Hausheer. 1926. Concerning two options in dilution egg counting: small drop and displacement. Am. J. Hyg. 6: 134–145.

Taylor, A.E.R. and J.R. Baker. 1968. The Cultivation of Parasites *In Vitro*. Blackwell Scientific Publications Ltd., Oxford, United Kingdom.

Truant, A.L., S.H. Elliott, M.Y. Kelley and J.H. Smith. 1981. Comparison of formalin-ethyl ether sedimentation, formalin-ethyl acetate sedimentation, and zinc sulfate floatation techniques for detection of intestinal parasites. J. Clin. Microbiol. 13: 882–884.

Van Dam, A.P., T. van Gool, J.C.F.M. Wetsteyn and J. Dankert. 1999. Tick-borne relapsing fever imported from West Africa: Diagnosis by quantitative buffy coat analysis and *in vitro* culture of *Borrelia crocidurae*. J. Clin. Microbiol. 37: 2027–2030.

Visvesvara, G.S. 2002. *In vitro* cultivation of microsporidia of clinical importance. Clin. Microbiol. Rev. 15: 401–413.

Visvesvara, G.S. and L.S. Garcia. 2002. Culture of protozoan parasites. Clin. Microbiol. Rev. 15: 327–328.

Wang, L.C. 1998. Evaluation of quantitative buffy coat analysis in the detection of canine *Dirofilaria immitis* infection: a model to determine its effectiveness in the diagnosis of human filariasis. Parasitol. Res. 84: 246–248.

Watson, J.M. and R. Al-Hafidh. 1957. A modification of the Baermann funnel technique and its use in establishing the infection potential of human hookworm carriers. Ann. Trop. Med. Parasitol. 41: 15–16.

Yang, J. and T. Scholten. 1977. A fixative for intestinal parasites permitting the use of concentration and permanent staining procedures. Am. J. Clin. Pathol. 67: 300–304.

CHAPTER 9

Interpreting Routine Microbiology Results for Patient Care

Shivanjali Shankaran

Despite vaccination, improvements in sanitation and the development of antibiotics, infectious diseases continue to account for a significant percentage of outpatient and inpatient visits. A study of geriatric patients showed that 13.5% of emergency room visits were for infectious diseases with 57% of these patients being admitted to the hospital (Goto et al. 2016). The management of patients is complicated by antibiotic resistance and emerging and reemerging infectious diseases. In this setting, it is essential for the clinician to know what tests to run, and how to interpret the results of those tests.

Routine microbiology tests as below provide vital assistance to the clinician in the evaluation of infections.

Test	Example
Gram stain and Cultures	Blood, urine, sputum, abscess, cerebrospinal fluid
Antigen testing	Pneumococcal antigen, Legionella antigen
Molecular testing	Gonorrhea and Chlamydia Nucleic acid amplification test
Serology	Syphilis testing, Hepatitis testing

Bacteria, viruses, fungi and acid fast bacilli can be cultured from various body fluids including blood, sputum, urine, cerebrospinal fluid, etc. Culture data guide antibiotic choice and duration; additional information such as prior antibiotic use can help with better interpretation of these

Assistant Professor, Division of Infectious Diseases, Department of Internal Medicine, Eastern Virginia Medical School, 825 Fairfax Avenue, Ste 410, Norfolk, VA 23507, USA.
Email: shankas@evms.edu

results. Identification of antigens provides a faster way of identifying pathogenic organisms. Knowledge of their sensitivity and specificity assist the clinician in giving appropriate weight to the test results. Molecular and serological testing are additional methods clinicians use routinely in the diagnosis of infectious diseases. The below discussion expands further on the interpretation of the above microbiology tests.

Interpreting Gram Stain and Culture Results

> A 72-year-old-woman is sent in from dialysis after she developed hypotension, chills and fever during her dialysis session. Due to the presence of her dialysis catheter, blood cultures are sent and she is started on Vancomycin to cover gram positive skin flora. Twelve hours later, blood culture gram stain shows a gram negative rod and Piperacillin-tazobactam is added to her regimen. Forty-eight hours later the bacteria is identified as *Proteus mirabilis* susceptible to multiple antibiotics. The positive gram stain allowed the clinician to add an appropriate antibiotic 48 hours before culture results were available.

Gram stains may provide invaluable antibiotic guidance to the clinician. For example, the presence of gram positive cocci in chains in a blood culture suggest a Streptococcus or Enterococcus species. Similarly, the presence of gram negative rods may prompt the use of appropriate antibiotics hours or even days before the organism is identified. A study of almost 6,000 gram stains from blood cultures showed almost 100% sensitivity of blood culture gram stain when it came to gram positive cocci and gram negative rods (Søgaard et al. 2007). This does not hold true across all biological samples, however. In prosthetic joint infections, the sensitivity of gram stains can be as low as 19% (Spangehl et al. 2007). By contrast, multiple studies have shown that the specificity of the gram stain is extremely high and surgeons may use a positive result to change the type of surgery they have planned (Oethinger et al. 2011). Similarly, gram stains of sputum samples have high specificity but lower sensitivity, and both depend on the quality of the sample collected (Rosón et al. 2003). As always, if a gram stain does not make sense, a discussion with the microbiologist can help to rule out contamination or false positive results. A positive culture, such as the one in the example above, provides a diagnosis and guides further investigations if needed. The identification of a *Proteus* species would suggest either a gastrointestinal or genitourinary source of infection. This in turn would guide appropriate management of the patient. While a negative culture could rule out an infection, the prior use of antibiotics as well as inadequate sample material could lead to a false negative result.

> A 34-year-old female with diabetes is admitted with community acquired pneumonia. She is started on antibiotics and a sputum culture is sent. An infectious diseases consultant is called due to the presence of *Neisseria* species in her culture. She has been informed that she may have gonorrhea and is extremely upset about this.

A knowledge of the normal flora of different body systems can help to identify true pathogens versus colonizers. In the example above, the *Neisseria* species are members of the normal oral flora (Aas et al. 2005), and do not need to be treated. This also indicates that the sputum sample was not appropriately collected. Similarly, cultures can get routinely contaminated if appropriate aseptic precautions are not followed. The identification of certain organisms, such as Coagulase negative *Staphylococcus*, non anthracis *Bacillus* species or *Propionibacterium* or their presence in just a single bottle, or set, typically points to contamination. The presence of *Pseudomonas aeruginosa*, *Staphylococcus aureus* or *Candida albicans* in blood cultures is presumed to be invariably pathogenic (Weinstein et al. 1997).

Viral cultures can take much longer to return and in these situations, the decision may be made to treat for a presumed infection based on risk factors and presentation. Alternate testing is required to establish a diagnosis in these situations. One example of this is Cytomegalovirus (CMV) disease in solid organ transplant recipients. CMV blood cultures are rarely useful due to their poor sensitivity. A CMV Qualitative Nucleic Acid Test (QNAT) from serum is helpful in diagnosing and monitoring disease in these patients. Immunohistochemical staining in pathology specimens can help to diagnose tissue invasion (Kotton et al. 2013). The simple presence of CMV DNA in blood, however, does not always indicate active CMV infection, implying that the clinical scenario guides how we interpret results. In patients with AIDS, a positive QNAT for CMV does not always reflect end organ disease. In an ACTG trial, 20% of patients were found to be CMV viremic, with only 5.8% of those having end organ disease (Wohl et al. 2009).

The Microbiology Sample

> A 52-year-old man comes in for routine evaluation. Over the last few months, he has noted increased urinary frequency. He denies any dysuria, change in color or smell of his urine or any abdominal pain. A urine culture is collected and returns with 50,000 colony forming units of *Klebsiella pneumoniae*. He is placed on a 7-day course of Ciprofloxacin without any resolution of symptoms.

To understand and appropriately treat a patient based on cultures, it is imperative to know where and how the sample was obtained. Appropriate sterile techniques must be utilized. Contaminated blood cultures are a good example of incorrect collection techniques leading to positive cultures, unnecessary antibiotic use and potentially longer hospital stays (Hall and Lyman 2006). Other diagnoses depend on volume, such as in endocarditis when multiple blood cultures are preferred. Similarly, when sending a sample from an abscess, aspirate is preferred over a swab due to higher number of organisms captured (Baron et al. 2013). When collection depends entirely on the patient, it is even more essential to ensure correct techniques. For example, discussing that lower respiratory tract sample is required and not just saliva when requesting a sputum sample. When obtaining AFB smears, the patient should not gargle with tap water as contaminating mycobacteria can give false positive results. Some tests should always be ordered together. A urine culture should be obtained with a urinalysis so that patients are not unnecessarily treated with antibiotics. In the situation above, a negative urinalysis may have helped confirm that this patient likely had asymptomatic bacteriuria, and limited antibiotic use.

Antigen Testing

> A 54-year old man with untreated AIDS is admitted with fever, malaise and a headache. A serum cryptococcal antigen returns positive and he undergoes a lumbar puncture which reveals elevated opening pressure. India ink staining is positive for encapsulated yeast and the patient is started on treatment for Cryptococcal meningitis. 8 days later, *Cryptococcus neoformans* is identified from CSF fungal cultures.

Antigen testing can provide a quicker way of diagnosing several infectious diseases. As in the case above, Cryptococcal capsular polysaccharide can be detected in various body fluids with high sensitivity and specificity (Feldmesser et al. 1996). This result can be obtained much earlier than culture data.

Primary care and hospital based clinicians routinely use antigen testing in the diagnosis of community acquired pneumonia (CAP). The pneumococcal urinary antigen test and the Legionella urinary antigen test are typically used. The former identifies the presence of capsular polysaccharide and has a sensitivity ranging from 50–80% (Mandell et al. 2007) with a specificity of up to 97% (Sinclair et al. 2013); one study revealed a sensitivity of just 64.5% in patients with pneumococcal bacteremia (Selickman et al. 2010). It is therefore important to keep these numbers in mind when using these tests to diagnose or treat patients with CAP. While the Legionella urinary antigen has high sensitivity and specificity, it only identifies soluble antigens from *Legionella pneumophila* serogroup 1. While this serogroup causes a majority of cases, this could be a potential weakness of the test. Additionally, this test can stay positive for many weeks after initial infection, therefore repeat positive results should be interpreted with caution (Reller et al. 2003).

Molecular Testing

> A 59-year-old man with poorly controlled diabetes is admitted with fever, confusion and weakness in his right arm for the last few days. A CSF analysis shows lymphocytic pleocytosis with mild elevation of protein. An HSV PCR in the CSF returns positive and he is started on Acyclovir. His confusion improves over the next few days.

The identification of pathogen RNA or DNA in serum or other body fluids provides an additional methodology in the diagnosis of infectious diseases. HSV Polymerase Chain Reaction (PCR) can be performed on multiple sites including CSF, skin vesicles and ulcers (Scoular et al. 2002). In these situations, cultures have lower sensitivities and may be negative with older lesions or in recurrences (Workowski and Bolan 2015). As PCR provides typing of HSV, it can assist with appropriate treatment, prevention and counseling of patients.

Nucleic acid amplification tests (NAAT) are routinely used in the diagnosis of sexually transmitted infections such as Gonorrhea and Chlamydia. Though only FDA approved for use with urogenital specimens (urine, urethral, vaginal or cervical swabs), several laboratories can run NAATs using rectal and oropharyngeal swabs, which can assist with clinical management. As these tests can identify the nuclear material of nonviable organisms, repeat testing too soon may give a false positive result. In case of suspected treatment failure or resistance, such as with gonorrhea, cultures should be obtained along with antimicrobial susceptibility (Workowski and Bolan 2015) as this information is not provided by NAAT testing alone.

Antibody Testing

When it is too difficult to isolate an organism, antibody testing may be used to establish a diagnosis. Clinicians can also use serology to assess need for and response to vaccination. It is important to know that a positive serologic test by itself does not establish a diagnosis of an active infection. In the diagnosis of syphilis, a treponemal test (such as Syphilis IGG) must be used in combination with a nontreponemal test (such as Rapid Plasma Reagin, RPR) to identify active syphilis vs. presence of antibodies due to prior exposure and treatment (Workowski and Bolan 2015). Additionally, a positive IGG may persist for life as happens after infectious mononucleosis with a positive EBV VCA IGG. Repeatedly testing for this or evaluation by a specialist for high titers is not typically warranted. As with all other microbiology testing, the results of serology testing must be evaluated in the context of patient symptoms and physical examination findings. When serology is used in acute infections such as certain tick borne and viral infections, a 4-fold increase in titers in convalescent serum can help ascertain the diagnosis (Baron et al. 2013).

Conclusion

Numerous tests can be used routinely in the diagnosis of infectious diseases. The choice of testing as well as evaluation of results depends on the sample drawn, the conditions in which it was collected as well as the patient presentation. Newer methods of microbiology diagnosis such as the Matrix-assisted Laser Desorption Ionization-Time of Flight (MALDI-TOF) are likely to decrease the time to identification of pathogenic microbes and assist with early adjustment of antibiotics used in treatment of these infections. As discussed later in this book, a close working relationship with the microbiology laboratory can assist with the appropriate interpretation of results.

References

Aas, J.A., B.J. Paster, L.N. Stokes, I. Olsen and F.E. Dewhisrt. 2005. Defining the normal bacterial flora of the oral cavity. J. Clin. Microbiol. 43(11): 5721–5732.

Baron, E.J., J.M. Miller, M.P. Weinstein, S.R. Richter, P.H. Gilligan, R.B. Thomson Jr. et al. 2013. A Guide to Utilization of the Microbiology Laboratory for Diagnosis of Infectious Diseases: 2013 Recommendations by the Infectious Diseases Society of America (IDSA) and the American Society for Microbiology (ASM). Clinical Infectious Diseases Advance Access published July 10.

Feldmesser, M., C. Harris, S. Reichberg, S. Khan and A. Casadevall. Serum cryptococcal antigen in patients with AIDS. Clin. Infect. Dis. 23: 827–830.

Goto, T., K. Yoshida, Y. Tsugawa, C.A. Camarago and K. Hasegawa. 2016. Infectious disease—Related emergency department visits of elderly adults in the United States, 2011–2012. J. Am. Geriatr. Soc. 64(1): 31–36.

Hall, K.K. and J.A. Lyman. 2006. Updated review of blood culture contamination. Clin. Microbiol. Rev. 19(4): 788–802.

Kotton, C.N., D. Kumar, A.M. Caliendo, A. Asberg, S. Chow, L. Danziger-Isakov et al. 2013. Updated International Consensus Guidelines on the Management of Cytomegalovirus in Solid-Organ Transplantation. Transplantation. Volume 96, Number 4.

Mandell, L.A., R.G. Wunderink, A. Anzueto, J.G. Bartlett, G.D. Campbell, N.C. Dean et al. 2007. Infectious diseases society of America/American thoracic society consensus guidelines on the management of community-acquired pneumonia in adults. Clinical Infectious Diseases 44: S27–72.

Oethinger, M., D.K. Warner, S.A. Schindler, H. Kobayashi and T.W. Bauer. 2011. Diagnosing periprosthetic infection: False-positive intraoperative gram stains. Clin. Orthop. Relat. Res. 469(4): 954–960.

Reller, L.B., M.P. Weinstein and D.R. Murdoch. 2003. Diagnosis of *Legionella* infection. Clin. Infect. Dis. 36(1): 64–69.

Rosón, B., J. Carratala, R. Verdaguer, J. Dorca, F. Manresa and F. Gudiol. 2003. Prospective study of the usefulness of sputum gram stain in the initial approach to community-acquired pneumonia requiring hospitalization. Clin. Infect. Dis. 31(4): 869–874.

Scoular, A., G. Gillespie and W. Carman. 2002. Polymerase chain reaction for diagnosis of genital herpes in a genitourinary medicine clinic. Sex Transm. Infect. 78(1): 21–25.

Selickman, J., M. Paxos, T.M. File, R. Seltzer and H. Bonilla. 2010. Performance measure of urinary antigen in patients with *Streptococcus pneumoniae* bacteremia. Diagn. Microbiol. Infect. Dis. 67(2): 129–33.

Sinclair, A., X. Xie, M. Teltscher and N. Dendukuri. 2013. Systematic review and meta-analysis of a urine-based pneumococcal antigen test for diagnosis of community-acquired pneumonia caused by *Streptococcus pneumoniae*. J. Clin. Microbiol. 51(7): 2303–2310.

Søgaard, M., M. Nørgaard and H.C. Schønheyder. 2007. First notification of positive blood cultures and the high accuracy of the gram stain report. J. Clin. Microbiol. 45(4): 1113–1117.

Spangehl, M.J., E. Mastrson, B.A. Masri, J.X. O'Connell and C.P. Duncan. 1994. The role of intraoperative gram stain in the diagnosis of infection during revision total hip arthroplasty. J. Arthroplasty. 14(8): 952–956.

Weinstein, M.P., M.L. Towns, S.M. Quartey, S. Mirrett, L.G. Reimer, G. Parmigiani et al. 1997. The clinical significance of positive blood cultures in the 1990s: A prospective comprehensive evaluation of the microbiology, epidemiology, and outcome of bacteremia and fungemia in adults. Clin. Infect. Dis. 24(4): 584–602.

Wohl, D.A., M.A. Kendall, J. Anderson, C. Crumpacker, S.A. Spector, J. Fenberg et al. 2009. Low rate of CMV end-organ disease in HIV-infected patients despite low CD4+ cell counts and CMV viremia: Results of ACTG protocol A5030. HIV Clin. Trials. 10(3): 143–152.

Workowski, K.A. and G.A. Bolan. 2015. Sexually transmitted diseases treatment guidelines. 2015. MMWR Recommendations and Reports/Vol. 64/No. 3.

CHAPTER 10

The Pharmacist Role in Antimicrobial Stewardship and Interpreting Microbiology Laboratory Results

Stephanie Crosby,[1,a] *Mark DeAngelo*[1,b] and *Nancy Khardori*[2,*]

BRIEF BACKGROUND ON HISTORY OF PHARMACY AND ANTIMICROBIAL STEWARDSHIP

Pharmacists have not always been an integral part of the antimicrobial stewardship (AMS) team. The transition has been a more recent development beginning with the education requirement changes in the pharmacy education program. Until 2003, pharmacists were only required to obtain a Bachelor's degree to practice, with the title of Registered Pharmacist (RPh). In 1997, the Accreditation Council for Pharmacy Education (ACPE) recommended changes to the degree, ultimately requiring any pharmacist graduating after 2003 to obtain a Doctorate of Pharmacy (Pharm.D.) (ACPE 2006, 2011). Since that time, the profession has markedly grown, offering post-graduation residencies, fellowships, board certifications and other specializations in various scopes of practice.

[1] DePaul Medical Center, 150 Kingsley Lane, Norfolk VA 23505.
[a] Email: Stephanie_Crosby@bshsi.org
[b] Email: Mark_DeAngelo@bshsi.org
[2] Professor of Medicine and Microbiology and, Molecular Cell Biology, Division Director, Infectious Diseases, Department of Internal Medicine, Eastern Virginia Medical School, Norfolk Virginia.
[*] Corresponding author: khardoNM@evms.edu

With the field of pharmacy becoming more involved in clinical aspects of patient care, especially in hospitals and other healthcare institutions, many organizations are utilizing pharmacists as a part of multidisciplinary teams, including antimicrobial stewardship programs. According to the Infectious Disease Society of America (IDSA), antimicrobial stewardship should focus on optimizing patient outcomes while limiting antimicrobial use to prevent toxicity and development of resistance (Delit et al. 2007). The antimicrobial stewardship program should include both infectious-disease trained physicians and pharmacists as the core of the team, with other participants including microbiologists, infection control specialists, informatics and epidemiologists if available (Patel and MacDougall 2010).

The American Society of Health-System Pharmacists (ASHP) has taken the position that pharmacists should have prominent roles in antimicrobial stewardship and infection prevention in health care systems (ASHP 2010). Pharmacists have been shown to be invaluable members of the team, offering patient-centered therapy as well as potential cost savings. For example, pharmacists look for opportunities to convert from parenteral to oral therapy when possible, recommend de-escalation of care based on microbiology results, offer pharmacokinetic monitoring of antibiotics, and assist with *Clostridium difficile* infection sparing regimens.

The degree of pharmacist involvement on the team can differ substantially depending on the area of practice. For example, large university hospitals will typically have a dedicated antimicrobial stewardship team with a pharmacist fully integrated in the role. This pharmacist will likely be a specialized pharmacist, having 2 years of intensive post-graduate training in infectious diseases. Some of these larger institutions may have daily rounds with a team consisting of physicians, pharmacists and microbiology representatives (MacVane et al. 2016). An example of an intervention made on interdisciplinary rounds is to request susceptibility testing on additional antimicrobial agents based on local prevalent resistance patterns or if alternate therapeutic options are needed due to allergies or other adverse events in a given patient.

Smaller community hospitals might not have a full team dedicated to antimicrobial stewardship on a regular basis, but the team might have a physician champion working with pharmacists to develop clinical programs that allow the pharmacy team to practice more independently (Patel and MacDougall 2010). Some interventions made by the AMS team, regardless of size, include maintaining a formulary of antibiotics, and enforcing restrictions on certain agents to limit over-utilization in unnecessary circumstances. The pharmacists may participate in other select services, including parenteral to oral conversion of select antimicrobial agents and assist in pharmacokinetic, e.g., renal dosing of antimicrobial agents.

Initiating Antimicrobial Therapy

Deciding to start a patient on antibiotics is ultimately up to the medical provider, but pharmacists can assist in the antimicrobial regimen of choice using basic principles of antimicrobial stewardship. Typical signs of infection include an elevated white blood count, with normal range being anywhere from 4500 to 10000 cells/mm^3 (Rybak and Aeschlimann 2008). When the patient is fighting an infection, neutrophils, consisting of both mature neutrophils known as segmented cells and immature neutrophils known as bands, will begin to escalate. The increase in bands is known as 'left shift'. Pharmacists look for this 'left shift' on the complete blood count to assist in differentiating between infection and other causes of leukocytosis, such as steroid use. Other laboratory values that are useful are lactic acid and procalcitonin, which are usually elevated in patients with true infections. Additionally, pharmacists will look for other signs of infection including sputum production, drainage, swollen areas or redness, and temperature changes from baseline.

In order to determine the appropriate initial therapy, the provider and the pharmacist should first be familiar with the institution's antibiogram. This tool reports susceptibilities from the previous year within a particular institution and surrounding areas to various antibiotics, empowering the clinician to choose the right presumptive therapy. Typically, the institution's microbiology laboratory will supply the susceptibility data and will work closely with the infectious disease pharmacist to prepare and distribute the antibiogram. If a gram negative infection is suspected in the hospital setting, common practice is to provide double coverage for *Pseudomonas aeruginosa* with an anti-pseudomonal beta-lactam, e.g., a carbapenem and either an aminoglycoside or a fluoroquinolone. The pharmacist will recommend to the provider to choose the two most appropriate agents by utilizing the antibiogram. For example, an institution might report 70% sensitivity rate to fluoroquinolones for *Pseudomonas aeruginosa*, whereas carbapenem or aminoglycosides have sensitivities > 90%. This would illustrate that fluoroquinolones might not be the best initial agent if *P. aeruginosa* is suspected, as with ventilator-associated pneumonia or septic shock. Moreover, the antibiogram is useful in preparing order-sets to assist the clinician with initial regimens for common infections.

Using Patient-Specific Factors When Initiating Therapy

A major aspect of determining initial therapy is patient-specific factors, such as medication allergies. It would be imprudent of the pharmacist to recommend a therapy that would be harmful to the patient. There are certain circumstances, however, in which a patient might need to be challenged with an antibiotic if it is the only option for the infection being managed. For example, if a patient has a true anaphylactic reaction to the penicillin class, but requires a penicillin-based therapy, the pharmacist could recommend a desensitization approach. In these instances, the patient is given small measured, increasing doses of the challenging agent over time under close watch by a team that can intervene appropriately if the patient shows signs of a hypersensitivity reaction. Most times, however, the patient might report "rash" from a penicillin agent, which would allow a challenge of other Beta Lactam therapies such as the cephalosporin or carbapenem classes if these agents were warranted. The cross-reactivity between penicillins and cephalosporins was found to be overly reported in early studies (10%) with actual cross-reactivity being closer to 1% (Campagna et al. 2012). Another reaction reported as an allergy is redness from vancomycin. This could be "Redman syndrome", a histamine released reaction which can be prevented in the future by doubling infusion time.

Aside from allergies, other patient conditions need to be taken into account before initiating antimicrobial therapy. For example, if the patient has a seizure disorder, the pharmacist would discourage use of some carbapenems. Patients with cardiac defects with prolonged QTC interval should not be exposed to further prolonging agents, such as fluoroquinolones, especially if other QTC prolonging agents are active on the patient's medication list. Patients with acute renal failure should not be initiated on nephrotoxic agents, such as aminoglycosides, unless other agents are not possible due to resistance patterns. On the other hand, if the patient has very good kidney function, the dosing needs to be initiated on the more frequent side of the recommended dosing range for optimal outcomes. Other considerations to keep in mind include, but are not limited to, pregnancy, lactation and hepatic insufficiency.

Another factor pharmacists can utilize in assisting with initial therapy is previous microbiology information if provided in the patient record. While not all infections recur, a patient with a history of multi-drug resistant organisms might warrant therapy that accounts for that resistance pattern to prevent treatment failure. For example, if a patient has grown enterococcus in previous urine culture, initiating a therapy with ampicillin would be better than a cephalosporin until new culture result is available. Patients with a history of methicillin-resistant *Staphylococcus aureus* (MRSA) should be given vancomycin pre-operatively instead of cefazolin in the setting of orthopedic procedures or other procedures requiring hardware placement. The pharmacist will always want to review the patient for any previous history of *Clostridium difficile infection (CDI)*. For patients with *CDI* in the past, the pharmacist might recommend to limit broad spectrum therapy, and to avoid agents more likely to cause *CDI* such as clindamycin. In patients with suspected *CDI* or recurring infections, the pharmacist will also consider discontinuation of a proton pump inhibitor if that was chosen for stress ulcer prophylaxis.

The site of infection is also a determinant in choosing initial agent(s). Certain antibiotics may not penetrate the site of infection, and should therefore not be recommended. Fluoroquinolones are often prescribed for urinary tract infections, but while ciprofloxacin and levofloxacin are excreted in the urine, moxifloxacin is not, so moxifloxacin would not be an appropriate agent. If a central nervous system infection is suspected, the blood brain barrier may limit what antibiotics can penetrate. Vancomycin and ceftriaxone will cross the blood brain barrier, but cefazolin will not (Leekha et al. 2011). If a patient has suspected MRSA pneumonia, daptomycin should not be initiated since it is inactivated by surfactants in the lungs.

Utilizing Other Strategies for Better Patient Outcomes

The mechanism of action or activity of the agent being initiated also needs to be considered. Some antibiotics are bactericidal whereas some may be bacteriostatic. Bactericidal agents will kill the bacteria by disrupting the cell membrane, cell wall or the cellular DNA (Nemeth et al. 2015). Examples of antibiotics in the category include beta-lactams, fluoroquinolones and glycopeptides. Agents that prevent further bacterial replication versus killing the organism are considered bacteriostatic. Antibiotics that are bacteriostatic include macrolides and lincosamides. In severe infections, such as meningitis or endocarditis a bacteriostatic agent would not be ideal as more aggressive killing would favor a better clinical outcome.

Pharmacists can take advantage of the pharmacodynamics and pharmacokinetics of agents to maximize efficacy of the drug. Examples include utilizing the post-antibiotic effect of agents by giving high doses spaced out. This can be done with aminoglycosides to maximize the concentration over the minimum inhibitory concentration, while accounting for the post-antibiotic effect of the drug. Patients on an extended-interval therapy will receive a large dose every 24, 36 or 48 hours depending on a 10 hour post-dose level compared to a nomogram. This approach is not recommended currently in endocarditis where gentamicin is used for synergy only because it has not been studied in that setting.

Beta-lactams have been shown to be more effective if the infusion time is increased to a 3 or 4 hours versus a 30 minute infusion. This strategy utilizes a functionality of time above the area under the curve (AUC) killing. This has shown promise in reducing length of stay and decreasing mortality (Nagel et al. 2014). Pharmacists should evaluate patients for appropriateness of extended beta lactam infusions and convert them to such therapy when possible.

Once the initial regimen is recommended, the pharmacist should adjust the dose appropriately for the type of infection, as well as renal or hepatic function of the patient. Some antibiotics like levofloxacin have different dosing regimens for various types of infection to be more efficacious. A dose for community acquired pneumonia would be 500 mg every 24 hours, whereas a dose for hospital acquired pneumonia (where *P. aeruginosa* is a possibility) would be 750 mg every 24 hours.

In addition to the dose, the route of therapy needs consideration. Severe infections typically warrant parenteral therapy, whereas some infections can be treated with oral therapy, although more likely in the outpatient setting.

The initial drugs should also be formulary agents of choice for the institution in addition to being broad spectrum in coverage against potential pathogens until more information is available from the microbiology laboratory.

Using the Microbiology Lab Clues to Modify Therapy

Microbiologists provide useful information early on in the pathogen identification process, which can help narrow the spectrum of therapy (de-escalation) or in some cases, indicate escalation of coverage. In most cases, results of the cultures are not available for 48 to 72 hours. The earliest tool utilized in identification of the bacteria is the gram stain, which is a technique that differentiates between gram positive and gram negative bacteria by creating differential color visualization (Rybak and Aeschlimann 2008). Gram positive bacteria will show as purple under the microscope, whereas gram negatives show as pink. The microbiologist will also see the shape of the bacteria, and can report if they are cocci in clusters or chains or bacilli. Pharmacists can utilize this data to start eliminating bacteria and help streamline therapy. For example, if gram positive cocci are the only bacteria seen on gram stain, the pharmacist should recommend discontinuation of gram negative coverage before the exact organism is even known. The cocci might clue the pharmacist to ensure there is coverage for Methicillin Resistant *Staphylococcus aureus* (MRSA) while awaiting identification and susceptibility results, especially if the institution has high MRSA rates.

The microbiologist can also report other features of the bacteria that help in the early identification process. If the bacilli are gram negative and non-lactose fermenting, this could help the pharmacist consider *Pseudomonas aeruginosa*, a very virulent bacterium that responds to a limited number of antibiotics. If the patient is on cephalexin for a urinary tract infection, appropriate intervention would be to escalate the therapy to piperacillin-tazobactam, meropenem, ciprofloxacin or cefepime. The choice of agent would depend on several patient factors, such as allergies, QTC prolongation issues, and even formulary agents in the hospital. Since it might take up to five days for the final culture and sensitivity report, the pharmacist can trend white blood cell count or temperature to monitor response.

When to Escalate

When a patient is on broad spectrum agents, there may be circumstances requiring further escalation of therapy. If the patient continues to have an elevated white blood cell count, continues to be febrile, or is hypotensive with tachycardia, the clinical picture shows that the patient is not responding to the current therapy and that a change is needed. One possibility is that the patient may have *Clostridium difficile* infection, which requires specific therapy to treat, such as metronidazole, (oral or intravenous), vancomycin (oral or enema) or fidaxomicin. If the patient is having diarrhea, it is prudent to collect a stool sample to test for *C. difficile* toxin and treat presumptively with metronidazole.

Another possibility is that the patient might have infection due to a multi-drug resistant organism. If the laboratory reports multiple antibiotic resistances, the pharmacist can assist the team by soliciting extended sensitivities to agents not commonly reported due to formulary restrictions or methodology used. Colistin testing is added for multi-drug resistant infections, but it might take several days to obtain results. In the meantime, if there are no other options, the pharmacist could recommend initiating colistin until the results come back for bacteria like multi-drug resistant *Acinetobacter*. Most laboratories do offer alternative testing of agents if the bacteria are resistant to the agents in the initial panel. The laboratory way need to provide extended/alternate sensitivity testing even in the absence of resistance to multiple agents due to patient factors like allergies and co-morbid conditions.

Another clue that the microbiology laboratory can provide is the speciation of the microorganism, in the absence of susceptibility testing. This is particularly helpful with microorganisms that have inherent mechanisms of resistance. For example, if the laboratory reports a *Candida* species is growing in the blood, typically fluconazole is started. If that species isolated is *Candida glabrata*, the pharmacist can recommend a change from fluconazole to micafungin, as *C. glabrata* has inherent resistance to azoles. *In vitro* susceptibility testing is not done routinely for fungal organisms, but understanding inherent resistance patterns will enable the pharmacist to recommend appropriate intervention.

Interpreting *In Vitro* Susceptibility Results

The microbiology laboratory reports the final culture and susceptibility, along with minimum inhibitory concentration (MIC) of the antibiotics against the organism. This allows classification of the bacteria into susceptible, intermediate or resistant to the tested antibiotics. Susceptible indicates that the drug is able to inhibit the bacteria at a standard dose. Resistant microbes show higher MICs, meaning the maximal dosing of the agent would elicit minimal antibacterial response, if any (Rybak and Aeschlimann 2008). Intermediate may indicate to the practitioner to utilize a higher than standard dose in order to inhibit that bacteria. However, if the antibiotic concentrates in the area of infection, such as the urinary tract, due to clearance of the drug, an intermediate susceptibility may still allow standard dosing. It would not, however, work at a site where the antibiotic does not have extensive penetration.

Pharmacists also have to be aware that MICs are not comparable between antibiotics, as they are specific to the antibiotic and the type of bacteria being tested (Leekha et al. 2011). If the laboratory reports an MIC of 1 mg to drug "A" and an MIC of 2 mg to drug "B", that does not mean that drug "A" is twice as likely to kill the bacteria as drug "B". It is important to note that the microbiology laboratory will only use and report antibiotics that are expected to be effective against the bacteria, and with that, usually only one representative of each class of antibiotics is reported. For example, with enterococcus, cephalosporins are not tested for MICs since they are known to not have activity against enterococci. The MIC part of the culture and susceptibility report aids the pharmacist in making recommendations to limit therapy to the narrowest spectrum agent, chosen from the ones that show *in vitro* susceptibility against the bacteria tested.

Caveats to the *in vitro* susceptibility report include that results might suggest sensitivity to some agents that are ineffective in reality. This could be for a number of reasons, including site of infection or not being able to account for drugs that require additional therapy for efficacy. Rifampin is commonly shown in *Staphylococcus aureus* sensitivity reporting, but should never be used as monotherapy, as resistance can develop quickly (Perlroth et al. 2008). Additionally, moxifloxacin might appear in *Escherichia coli* sensitivities, but it is ineffective if the source is urine due to lack of renal excretion. Aminoglycosides show *in vitro* sensitivity but should not be used in meningitis due to lack of penetration into the cerebrospinal fluid.

A challenge to pharmacists can be misleading sensitivities due to complex mechanisms of resistance patterns in some of the multi-drug resistant pathogens. Some extended-spectrum beta-lactamase (ESBL) producing bacteria might show sensitivity to cefoxitin, but resistance to ceftriaxone

(Leekha et al. 2011). Further testing might be warranted to prove this is indeed an ESBL producing organism that would ultimately require a carbapenem. This has led some institutions to alter the reporting for cephalosporin so they do not appear falsely effective leading to poor clinical outcomes as cephalosporin are ineffective against ESBL bacteria (Rodloff et al. 2008).

Once the pharmacist has assisted in switching antibiotic therapy to the narrowest spectrum appropriate agent, the pharmacist should also recommend an end date so the therapy does not continue beyond its recommended course as antibiotics are associated with collateral damage the normal flora. In addition, antibiotics have side effects on the host, however minimal or severe they may be. Some therapies, such as colistin, will cause temporary renal insufficiency. Other side effects could range from gastrointestinal disturbances such as diarrhea to severe hematologic effects such as thrombocytopenia or hemolytic anemia. Another problem presented by antibiotics use is that of developing secondary infections. By modifying the normal flora especially in the gastrointestinal tract antibiotic may lead to new infections, such as candida infections or *Clostridium difficile* infections. These infections will then need to be treated as well. Therefore, by limiting the number of agents the patient is exposed to, as well as the length of exposure, the pharmacist can help prevent side effects, colonization and even new infections.

In addition to narrowing the antibiotics, the pharmacist should also look for opportunities to convert the parenteral therapy to oral as soon as possible for most infections. The exceptions are meningitis and endocarditis, where oral therapy might not provide adequate penetration into the site of infection (Leekha et al. 2011). The use of IV to PO conversion can decrease length of stay, decrease line associated infections, and ultimately decrease costs. The pharmacist should convert appropriately based on manufacturer recommendations, as not all antibiotics are 1:1 conversion between intravenous and oral formulations.

When Cultures are Not Diagnostic

There are times that no organism grows in a finalized culture. This could represent a multitude of things. Primarily, perhaps the patient does not have a bacterial infection, and antibiotics are unnecessary. This could indicate the infection is viral. Sometimes cultures are collected after initiation of antibiotic therapy, so the pharmacist should note time of cultures versus initiation of therapy. If cultures are collected after a few doses of antibiotic therapy, nothing may grow but discontinuation of therapy would not necessarily be appropriate. The patient might need to stay on broad spectrum antibiotics for the duration of therapy. Sometimes, microorganisms may grow on cultures that are not the causative pathogen. Candida species often appears in the sputum, but this represents colonization, and does not warrant treatment. It is the responsibility of the antimicrobial stewardship team and the provider to determine colonization versus true infection.

Coagulase-negative Staphylococcus (CONS) is a common contaminant that appears in blood cultures as a result of technique errors. When CONS grows in a single blood culture, the microbiology laboratory most likely will not report sensitivities. However, if the provider believes it to be a true pathogen due to the patient factors, further investigation may be done by repeating blood cultures. If the organism reappears with proper blood draw technique, a sensitivity request to the laboratory is justified. Other common contaminants include lactobacillus in urine cultures, if not collected as a clean catch specimen. Determining contaminants and treatment or lack thereof is an important role of the antimicrobial stewardship team.

Assessing the Interventions

Once the agent and length of therapy has been established, the pharmacist can help the healthcare team monitor the patient for response to the therapy. The white blood cell count and temperature trends should be tracked to ensure the patient is responding. Other follow up interventions could be radiologic findings on chest x-rays, repeat cultures, respiratory status improvement or blood pressure changes, depending on the initial infection. Other laboratory values, such as lactic acid and procalcitonin, can be trended in severe infections to curtail antibiotics if the patient is responding to therapy. As long as the patient is responding appropriately, the pharmacist might even be able to recommend oral therapy, depending on the site of infection. Transitioning the patient to oral therapy may allow the patient to be discharged earlier, which could further assist in cost savings from the AMS intervention.

At discharge, the pharmacist can assist the team by ensuring that oral therapeutic doses are appropriate. If the patient needs extended parenteral therapy for osteomyelitis or endocarditis, the pharmacist can help tailor it to once a day especially if the patient needs to come to an infusion center. If the patient's infection warrants parental therapy for the remainder of the course, the pharmacist can assist timing of antibiotic to facilitate infusion center times or home care visit times depending on the disposition of the patient. The pharmacist is able to educate the patient on timing of the medication, as well as any drug interactions to be considered. For example, certain iron supplements or multivitamins might interfere with absorption of the antibiotics like tetracyclines. An important counseling point should be dietary considerations, as some antibiotics should be taken on an empty stomach, whereas others require food. Gastric acid reducing medications like H2 blocker and proton pump inhibitors prevent the absorption of azoles like fluconazole.

Summary

Recently, the healthcare system has undergone reimbursement transformations, where several organizations can invoke financial penalties if certain quality metrics are not being met (Nagel et al. 2014). One method to achieve quality outcomes and reduce costs is the antimicrobial stewardship program. In addition to promoting appropriate therapy, AMS serves to decrease the number of multi-drug resistant organisms, as well as prevent harm to the patient by minimizing subsequent infections such as *Clostridium difficile* (MacVane et al. 2016). The work of pharmacists and the team in antimicrobial stewardship has been demonstrated to improve patient care, reduce medical costs, and reduce readmission rates, with the latter saving the organization further financial penalties as readmissions are penalized (Nagel et al. 2014). The Center for Medicare and Medicaid Services (CMS) now require reporting of hospital acquired *CDIs*, indicating more than ever the growing importance of a robust antimicrobial stewardship team to help prevent this super-infection (Nagel et al. 2014). The American Society of Health-System Pharmacists (ASHP) recommends that all pharmacists involved in acute care receive post-graduate training in residency programs as the use of pharmaceutical agents is becoming more complex (ASHP 2007). All of these reasons demonstrate the need for pharmacists to continue involvement in antimicrobial stewardship, using the clues provided by the microbiology laboratory to optimize patient care. This role for pharmacists is of particular significance and beneficial to patients in health care institutions not staffed by infectious diseases physician specialists.

References

ACPE Accreditation Standards and guidelines for the professional program in pharmacy leading to the Doctor of Pharmacy degree. Guidelines 2.0: January 23, 2011.

ASHP Statement on the Pharmacist's Role in Antimicrobial Stewardship and Infection Prevention and Control. 2010. Am. J. Health-Syst. Pharm. 67: 575–577.

American Society of Health-System Pharmacists. 2007. ASHP long-range vision for the pharmacy work force in hospitals and health systems. Am. J. Health-Syst. Pharm. 64: 1320–30.

Campagna, J.D., M.C. Bond, E. Schalbelman and B.D. Hayes. 2012. The use of cephalosporins in penicillin-allergic patients: a literature review. J. Emerg. Med. 42(5): 612–620.

Delit, T.H., R.C. Owens, J.E. McGowan, D.N. Gerding, R.A. Weinstein, J.P. Burke et al. 2007. Infectious disease society of America and the society for healthcare epidemiology of America guidelines for developing an institutional program to enhance antimicrobial stewardship. Clinical Infectious Diseases 44: 159–177.

Leekha, S., C.L. Terrell and R.S. Edson. 2011. General principles of antimicrobial therapy. Mayo Clin. Proc. 86(2): 156–67.

MacVane, S.H., J.M. Hurst and L.L. Steed. 2016. The role of antimicrobial stewardship in the clinical microbiology laboratory: Stepping up to the plate. Open Forum Infect. Dis. 3(4): ofw201.

Nagel, J.L., J.G. Stevenson, E.H. Eiland and K.S. Kaye. 2014. Demonstrating the value of antimicrobial stewardship programs to hospital administrators. Clinical Infectious Diseases 59(S3): S146–53.

Nemeth, J., G. Oesch and S. Kuster. 2015. Bacteriostatic vs. bactericidal antibiotics for patients with serious bacterial infections: systematic review and meta-analysis. J. Antimicrob. Chemother. 70(2): 382–395.

Patel, D. and C. MacDougall. 2010. How to make antimicrobial stewardship work: Practical considerations for hospitals of all sizes. Hosp. Pharm. 45(11 Suppl 1): S10–S18.

Perlroth, J., M. Kuo and J. Tan. 2008. Adjunctive use of rifampin for the treatment of *Staphylococcus aureus* infections: A systematic review of the literature. Arch. Intern. Med. 168(8): 805–819.

Rodloff, A., T. Bauer, S. Ewig, P. Kujath and M. Eckhardr. 2008. Susceptible, intermediate, and resistant—The intensity of antibiotic action. Dtsch. Arztebl. Int. 105(39): 657–62.

Rybak, M.J. and J.R. Aeschlimann. 2008. Laboratory tests to direct antimicrobial pharmacotherapy. pp. 1715–1730. *In*: DiPiro, J.T., R.L. Talbert, G.C. Yee, G.R. Matzke, B.G. Wells and L.M. Posey (eds.). Pharmacotherapy: A Pathophysiological Approach. McGraw Hill, New York, NY, USA.

CHAPTER 11

The Liaison and Collaborative Functions of the Clinical Microbiology Laboratory

Patrick G. Haggerty

INTRODUCTION

As many as 1 in every 25 hospital admissions acquires an infection (Magill et al. 2014) resulting in an estimated cost of over $9.8 billion annually in the United States (Zimlichman et al. 2013). It is imperative that our professionals who are tasked to discover, diagnose, manage and treat infections communicate often and work together to provide the most timely and accurate services to our patients. This collaboration enhances the work of the clinician and the microbiologist alike. The hospital benefits by timely and accurate treatment of patients, thereby improving outcomes and decreasing hospital length of stay. The system benefits from this collaboration by improving infection control, antibiotic stewardship, staff training, and process improvement. Due to costs involved, many hospital systems in the U.S. and Europe have attempted to outsource microbiology services to high through-put laboratories (Read et al. 2011, Murray and Witebsky 2009). This practice has a negative impact on communication between the clinician and the microbiologist and will ultimately lead to fragmented care and increased cost. Therefore, Peterson and colleagues have emphasized that "maintaining high-quality clinical microbiology laboratories on the site of the institution that they serve is the current best

Associate Professor, Division of Infectious Diseases, Department of Internal Medicine, Eastern Virginia Medical School, Norfolk, Virginia.

approach for managing today's problems of emerging infectious diseases and antimicrobial agent resistance" (Reller et al. 2001).

Sautter and Thompson go to great lengths describing the benefits and pitfalls for consolidating the microbiology laboratory to an off-site central testing area (Sautter et al. 2015). Thompson's most compelling rationale for keeping the laboratory at patient care sites is the development of relationships with care providers. In this manner the clinical microbiologist can educate the clinician as to the appropriate interpretation of results and actively participate in committees to oversee best practice. While tests may be run in a more cost effective manner with consolidation, if this communication is overlooked, inappropriate testing, misinterpretation of results, and unnecessary antibiotic prescription may result.

A. Acquiring and Ordering the Correct Specimen

Clinicians evaluating a patient with a potential infection should make certain that they order the right tests and submit the specimens via the right collection media. Saleem and Al-Surimi, in their process improvement efforts to reduce laboratory errors, found that most errors occurred in the pre-stage where the specimens are obtained, packaged, labeled and sent to the laboratory (Al Saleem and Al Surimi 2016). In the IDSA Guidelines on the Utilization of the Microbiology Laboratory, Baron et al. emphasize the need for a "close, positive working relationship between the physician and the microbiologists" (Baron et al. 2013). This will ensure that "selection, collection, transport, and storage of the patient specimens are performed properly" (Baron et al. 2013). Based on the clinical scenario, there should be a presumption of what kinds of organisms are likely to be present. The clinician should notify the laboratory of what is expected so that they can plate the specimen on the proper media. There may be some special methods that need to be employed in the laboratory such as special media (e.g., BCYE agar for Legionella), special environment (e.g., CO_2 chamber for anaerobes) or special conditions (e.g., lower temperature incubation for *M. marinum* or Yersinia).

Murray (Murray and Witebsky 2009) emphasizes that it is the clinician's responsibility to obtain sufficient quantity of the specimen being tested preferably prior to the administration of antibiotics. The correct container should be selected and it should be transported in a timely manner. Depending on surrogates for transport, such as busy nurses or overwhelmed transportation teams, a delay could negatively affect the viability of the organisms to be cultured. The microbiologist should make available instructions to assist the clinician in proper methods of specimen acquisition and transport. Guidelines for effective specimen management have been written by Baron et al. (Baron et al. 2013) that resulted from collaboration of the Infectious Diseases Society of America and the American Society of Microbiology. The laboratory should also put into place procedures to process the specimens in a timely manner and to inform the clinician if testing will be delayed because a test is only performed at a certain frequency.

Lack of attention to detail here could cause false negative results and inappropriate or inadequate treatment of the patient. Baron et al. describe what specimens are needed for different types of pathogens and underscore that specimen collection should be by or under the close supervision of the ordering physician. Many times a nurse or technician who may have limited knowledge in this area is "caught in the middle" between the physician and the laboratory, resulting in uninterpretable or misinterpreted results (Baron et al. 2013). On occasion, specialized media needs to be gathered by the laboratory so that appropriate tests can be run. This may necessitate a delay in specimen collection (if not critical) or prompt transportation of the collected specimen (if critical) to a specialty laboratory where the testing options are more robust. Schofield outlines detailed methods to prevent errors in the microbiology laboratory, by discussing which collection methods are acceptable for surgically acquired specimens and what are acceptable timeline for storage and transport (Shofield 2006). These methods should be reviewed by all clinicians who are in a position to collect specimens especially since the collection of these specimens is not without risk to the patient and are difficult to re-acquire if rejected by the clinical microbiologist (CM).

Needs of clinicians in specialty areas are often overlooked unless open communication precedes specimen acquisition (Mannis and Holland 2016, Ho et al. 2010). In an ophthalmology practice, care needs to be taken in specimen acquisition since soft tipped cotton applicators may be inferior to calcium alginate swabs that have been reported to produce a higher yield of bacterial organisms (Mannis and Holland 2016). Most microbiology laboratories will not routinely stock media for an infrequent pathogen such as acanthamoeba. By discussing this with the laboratory before specimen acquisition, the laboratory can acquire the appropriate media and therefore successfully identify this organism. If this communication does not take place then the specimen may be rejected or the pathogen may be sub optimally isolated because of a delay in plating. The otolaryngologist frequently submits culture swabs from the external ear canal in order to isolate a pathogen and obtain guidance on antibiotic selections. Many microbiologists are not aware that the preferred method of treatment for external otitis is with topical agents. Most of the antibiotics listed on typical antibiograms are systemic (oral or IV) antibiotics which will be of little use to the physician in treatment selections (Ho et al. 2010). Discussion with the microbiologists before specimen submission could possibly ameliorate this disconnected system.

B. When Specimens Should and Should not be Cultured

All too often specimens may be received by the laboratory that are not worthy of culture since the results may be misleading for the treating clinician. One common example of this is sputum specimens. Since collection of these samples is frequently poorly supervised (sometimes by an assistant with little knowledge of the importance of good collection methods), samples may be inadequate and reflect only oral secretions. By doing a gram stain on the specimen, the laboratory can determine the "quality" of the specimen. If there are large numbers of epithelial cells with few WBCs, it is appropriate for the laboratory to discard these specimens. Culturing flora from the mouth will lead to inappropriate or inaccurate antibiotic prescriptions that will not necessarily achieve the expected clinical result. Exceptions to this rule may include neutropenia or deep tracheal samples acquired from an endotracheal tube. Communication with the laboratory about these exceptions ahead of specimen submission will help avoid the laboratory discarding these specimens by protocol and consequently delaying the diagnosis and appropriate treatment. Another common example of misleading results are wound cultures. The clinician should culture only wounds that exhibit signs of infection. If a decubitus ulcer shows an open wound that is chronic, is not associated with surrounding erythema and the patient has no fever or elevated WBC count, then the wound should probably not be cultured routinely. Organisms will be present in any open wound but the likelihood of them being significant or worthy of treatment with systemic antibiotics is low. Purulent wound drainage, surrounding area of inflammation, fever and other systemic signs of sepsis make culturing and targeting specific bacteria more important. The significance of a positive culture from a fistulous tract is low as well, but may be somewhat higher if *Staphylococcus aureus* is isolated. Several organisms are routinely considered "contaminants" by the microbiologist if coming from superficial wound cultures. These include *Staphylococcus epidermidis*, diptheroids and corynebacteria. Most commonly they are reported as "normal skin flora" and not subjected to antibiotic sensitivity analysis. This may serve as a frustration to the clinician if the specimen was obtained by sterile means with the culture of deeper tissue or bone.

Georgiou et al. (Georgiou et al. 2011) reported that 97% of laboratory professionals felt that patient related clinical information had an impact on both specimen processing and interpretation. Their audit of handwritten requests showed that only 43% contained clinical information but demonstrated that use of computerized provider order entry (CPOE) more than doubles this transfer of information. Open communication with the laboratory about the clinical scenario helps the clinician and the microbiologist work in concert to provide meaningful actionable data so that the patient can be appropriately treated without delay. Additionally, these pretest discussions may help the clinician order the right tests at the right time or avoid submitting a culture all together. If underlying bone is suspected to be infected, it is important to acquire the specimen from the bone through a clean incision for best interpretation of the results. One common error is the acquisition of bone for culture via Cope needle directly through an open wound. In these cases the microbiology results likely represent the surface colonizing wound organisms which could mislead the clinician into using broad-spectrum antibiotics.

The microbiology laboratory frequently has protocols to reject specimens if they are determined to be unhelpful. Some examples of this are: repeated testing for *Clostridium difficile,* lack of proper specimen (solid stool for *C. difficile* or stool cultures), or lack of infective or inflammatory markers in the urinalysis before urine culturing. These tests will waste valuable laboratory resources but more importantly will lead to inappropriate treatment. If the clinician has questions about these protocols, (s)he should speak to the laboratory director in a nonconfrontational manner. Exceptions to these protocols will certainly be taken into consideration.

C. Rapid Notification of Clinician of Drug Resistant Organisms (DRO) that Require Isolation

The microbiologist is a valued member of the infection control team of the hospital. One important task of the microbiology team is the rapid identification and notification of treating clinicians of the presence of DRO. The sooner the clinicians are notified, the sooner the patient can be put on appropriate isolation precautions, which will help prevent lateral transmission of organisms in the hospital. While clinicians are frequently encouraged to presumptively isolate patients with possible DRO (e.g., patients presenting with apical pneumonia and symptoms of tuberculosis, patients with diarrhea after an antibiotic course), this isolation is often delayed until reports emerge from the laboratory. A protocol for rapid notification to the physician and/or the floor nurse by phone is an efficient way to relay this information. Even more efficient may be an electronic means via the EMR or secure texting to dedicated infection control personnel. A two way communication should be built into the system so that information receipt can be confirmed. The nurses are often authorized to place an order for isolation armed with this information however they may consult with the treating physician or the infection control nurse if there are other questions or mitigating circumstances.

As a member of the infection control team, the microbiologist is always looking for ways to make the diagnosis of infectious diseases (particularly contagious diseases) more rapidly. With the recent emergence of robust PCR panels that identify organisms like *Clostridium difficile* (CDI) from stool samples or methicillin resistant *Staphylococcus aureus* (MRSA) from blood, wound and sputum samples, these diagnoses can be made more rapidly. This results in more timely isolation of patients with these pathogens. Beavers and Wheeler (Beavers and Wheeler 2010) demonstrated how the use of PCR for joint fluid analysis in one case shortened the time to microbiologic diagnosis of the organism by 75 hours. This led to a narrower spectrum antibiotic prescription, decreased length of stay, significant cost savings and consequently a better outcome. Often these methods, while rapid, are more expensive. Having the microbiologist on the team to discuss the cost analysis and ultimate overall benefit to the system will go a long way in getting these new methods adopted.

While the microbiology team typically starts very early in the day, they frequently are not ready to post their final results in the electronic medical record until later in the morning. Read et al. (Read et al. 2011) noted that this could be after rounds so may not be seen until the following day by the treating clinician. In some institutions, the pharmacists or infection control personnel who are not restrained by rounding patterns are tasked with looking for bug/drug mismatches and calling the clinician to report these findings. Moore and Koerner experimented with 24 hour a day work patterns and clinician notification of positive blood cultures but found "no compelling evidence" that this practice improved patient outcomes (Moore and Koerner 2015). By contrast, Huang et al. (Huang et al. 2015) found "microbiology rounds" that comprised a daily meeting for one hour with a pharmacist, the clinical microbiologist and the ID physician improved efficiency of results reporting. There was more timely de-escalation of antibiotics resulting in a reduction of antibiotic costs. New methods such as notification of the clinician by secure email or text have been proposed (Beavers and Wheeler 2010) especially when the result could change the treatment course or affect the need for isolation. Two-way communication, confirming the receipt of this information, is obviously ideal and should be built into any notification scheme. Regardless of the system employed, if the microbiologist notices an unusual or unexpected pathogen, (s)he should not hesitate to notify the physician.

D. Microbiology Results as Warning and Epidemiologic Tools

While infectious disease physicians, intensivists and other physicians may see a concerning trend in emerging infectious diseases, these clinicians only see a subset of the cultures going into a busy hospital microbiology laboratory. The microbiologist has the opportunity to see all of the cultures coming from every floor in the hospital and the emergency department. Some large hospital laboratories also service other facilities such as surrounding nursing homes, long term acute care hospitals (LTACH), and physicians' offices. Therefore, the microbiology supervisor is in an ideal position to identify trends in emerging infections, including the emergence of resistant pathogens. With the microbiology supervisor's involvement and communication with the clinician and the infection control team, proactive protocols can be put in place to limit spread of these infections. On occasion, an outbreak investigation needs to take place to identify the source of these infections and ultimately control this. Pathogens that are unusual, such as anthrax or tularemia, can tip the microbiologist off to a more emergent situation such as a bioterrorism attack.

E. Production of Antibiogram Every Six Month

Over the last 10 years, more robust computer systems have allowed the microbiology laboratory to report the hospital antibiogram on a regular basis. Ideally, it can be reported every six months or even updated monthly with the trailing 12 month data. While this information is by its nature retrospective, it is designed to help the clinician select antibiotics that are most likely going to be effective. Most initial antibiotic selections are presumptive since the need for treatment typically precedes the availability of the cultures and sensitivities by 48–72 hours. One example of this is the use of ampicillin for *E. coli* urinary tract infections. Most hospitals are reporting 20–30% resistance to ampicillin, so by seeing this on the antibiogram, the clinician would likely select a drug that has a higher chance of success. These results should be reviewed with the infection control team on a biannual basis so emerging resistance trends can be appreciated and managed.

As previously noted, many hospital systems collect culture material from a large catchment area. Because the resistance patterns may be different within the hospital (i.e., ICU vs. surgical floor) and external to the hospital (general hospital vs. outpatient clinics), it is important to have an antibiogram generated from each facility and even each part of the facility in the catchment area. This will allow a better interpretation of data by the clinician practicing in certain area. Community trends should also be determined by reviewing consolidated data with the infection control team. Local and regional antibiograms have been shown to be more useful than national databases for selection of presumptive antibiotic therapy (Var et al. 2015).

F. Active Member of the Infection Control (IC) Team

Kalenic and Bidimer describe the essential role of the clinical microbiology (CM) in infection control and outline methods used to effectively integrate them into the team functions (Nishi and Hidaka 2016). Public reporting of hospital acquired infections (HAI) and third-party reimbursements based on infection quality indicators, have made collaboration between the infection prevention (IP) team and the CM even more relevant (Barenfanger et al. 2009). The CM can help with the accuracy of reporting by discarding contaminated wound cultures or respiratory cultures that show only epithelial cells and no WBCs on gram stain. Certain bacteria that are considered contaminants such as diphtheroid in wound cultures and one of two blood cultures with *S. epidermidis* should not be reported as positive in the public reporting of surgical site infections or central line related blood stream infections, respectively.

It is important for the IC team and the CM to discuss any new protocols designed to prevent and control multi-drug resistant organisms (MDRO). One example is the screening for nasal colonization with MRSA in the intensive care unit population. The microbiologist can assist in getting the most cost effective testing methods by balancing the cost with the timeliness of the report. While PCR methods may be more rapid in identifying MRSA, the hospital system may decide to go with the more cost effective method using chromagen agar. Without advance preparation, this protocol may fail because of lack of appropriate materials or sufficient personnel needed in the microbiology lab for testing and result notification.

Effective use of limited resources can result from good collaboration between the CM, the IC team and the clinician. Some examples of this have been discussed by Barenfanger et al. (Barenfanger et al. 2009). If initial cultures have verified an influenza A outbreak in the community, further testing of symptomatic patients could be limited to only hospitalized patients. If multiple samples are submitted for the diagnosis of *C. difficile* when the laboratory is testing with GDH antigen/toxin testing or PCR, then the laboratory should have the authorization to discard all but the first sample. The IC team should educate clinicians and the nursing staff about the rationale for this practice.

At the 2015 joint meeting of the Association for Professionals in Infection Control and the American Society of Microbiology, a focus group of infection preventionists (IP) and laboratorians, noted that a gap between the two departments remains (Spencer et al. 2016). Therefore, they developed a roadmap for improved communication: (1) Train together: the IPs at the early stages of their career should spend time in the microbiology lab to familiarize themselves with the logistics and processes behind each test. (2) Establish a regular meeting time between the IPs and laboratorians to ensure a consistent exchange of information. (3) Run the numbers: identify the isolation days, average daily bed cost, infection rates and CMS penalty data to estimate the cost of hospital acquired infections. This can then be compared to the cost of new technology which may be more expensive at the outset but mitigate overall costs in the long run. (4) Share the burden. Economic burdens of the increased cost of laboratory testing should be shared across departments such as laboratory, infection control, pharmacy and in some cases nursing. (5) Expand the committee. The members of the infection control committee should definitely include the chief microbiologist so that (s)he can share trends seen in the lab with a group of other dedicated professionals in attendance from nursing, hospital administration, housekeeping, regional public health and pharmacy.

The microbiology team can also help in outbreak investigation by helping the clinician understand the ecology of the organisms, thereby designing better protocols for investigation. One example of this is in an outbreak of legionella in the intensive care unit. By knowing that common sources of this pathogen include water, shower and faucet heads, appropriate environmental cultures can be obtained. Avoidance of inappropriate cultures of food, employees, bed sheets and the walls will save a lot of resources and prevent needless investigation. While the CM is the "expert" in culture methods of clinical samples, environmental sampling may be outside of the scope of their expertise and violate the rules of their accreditation. Laboratories that specialize in environmental sampling may have to be consulted to identify and process airborne or surface pathogens when deemed necessary for an investigation (Barenfanger et al. 2009).

G. Active Member of the Antibiotic Stewardship Committee

Read et al. (Read et al. 2011) noted that the collaboration between pharmacists, infectious disease physicians, and clinical microbiologists both inside and outside the hospital helps to promote the appropriate use of antibiotics, contain costs, and limit the development of antimicrobial resistance. Several previously noted examples indicating the role of the microbiologist in rejecting certain specimens based on associated data or the type of organism isolated, lead to good antibiotic stewardship. This helps to avoid the inappropriate use of antibiotics, when no infection was present or when the wrong organism was identified because of poor collection practices.

The microbiologists and infectious diseases clinicians should work in concert with the pharmacy to set up panels of antibiotics for testing against various bacteria. This should closely mirror the antibiotics that have been approved for these organisms by the hospital's pharmacy and therapeutics committee. Newer drugs, such as tigecycline and ceftaroline for example, may be tested but results suppressed by the microbiologist for the initial report in an effort to lead the clinician to the more cost effective and narrower spectrum antibiotics. While many clinicians criticize this practice as merely a cost saving maneuver by the pharmacy, it is, in fact, an effort to direct the clinician to the most appropriate antibiotic and to avoid the practice of using "the latest and the greatest" antimicrobial to hit the market. Pharmaceutical drug representatives attempt to influence the clinician in this direction. It is imperative for the microbiologist to inform the leadership, the intensivist/pulmonologist and infectious disease physicians of the suppression practice so that they can obtain broader sensitivity information on the more resistant organism on demand to help guide therapy in the more complex cases. In some cases the microbiologist has a secondary panel for DRO, that can be triggered and reported immediately. This practice helps to avoid delay in reporting vital information to the treating clinician since the timing of appropriate antibiotic administration is critical.

One area where microbiologists can work effectively with the antibiotic stewardship committee is in the area of false positive or "contaminated" blood cultures (FPBC). By identifying the source of these contaminants, which frequently come from the busy emergency departments (Self et al. 2013), additional monitoring and training can lead to a reduction of these confusing and very costly results. By reducing the FPBCs, the needless use of antibiotics, additional testing and increase length of stay can be avoided (Alahadi et al. 2011).

The microbiologist is frequently the hospital's resident expert on new technology. Use of this new technology can help identify offending organisms more rapidly and allow treatment to be targeted to the true offending organism in a timely manner. This can help the clinician avoid many days of multiple broad-spectrum antibiotics and replace them with a more targeted regimen. In some cases, these new technologies, as they replace older methods, eliminate the need for antibiotics altogether. One example of this could be a patient who presents with a bilateral pneumonia. While culture results may take days, a polymerase chain reaction (PCR) in the blood showing MRSA will direct the clinician to the right antibiotic from the outset. Subsequent culture results are used to confirm these results and provide *in vitro* antibiotic susceptibility data. Another example is the adoption of PCR for the testing of *C. difficile* in cases where the preliminary test shows the antigen positive and toxin negative. If the reflex PCR test is negative for toxin then antibiotic prescription and isolation can be completely avoided. Since PCR methods are expensive, laboratories are often reluctant to expand their panels when they feel that other methods (including culture and antigen testing) are sufficient. A multidisciplinary team which includes pharmacy, infection control, pathology, microbiology and infectious diseases can get these decisions out of the "silo" of one department and into the arena of a true efficacy as well as cost analysis. While the tests may be more expensive than traditional methods, if they produce results that are more accurate and/or more timely they will save the hospital money due to appropriate or decreased antibiotic prescriptions and frequently decreased length of stay also. Numerous other benefits from this technology are discussed in other chapters of this book.

H. Education of the Clinician and Hospital Staff

The microbiologist is an important member of the ongoing education team. While informal discussions with the clinician are very important, more formal didactic sessions at hospital grand rounds or informal sessions at medical and surgical practice committee meetings are important in helping to inform the physicians of new methodology being launched. Errors that could be avoided with enhanced clinician knowledge should also be relayed at these meetings since many physicians are unaware of the sensitivity and specificity of certain laboratory tests. If a physician orders a nasal swab for influenza and it is negative, it could be truly negative or it could be a false negative based on the 60% sensitivity of the test. If the clinician is made aware of this and the patient has a clinical syndrome compatible with an influenza-like illness, treatment may still be warranted. This information can be relayed at didactic sessions, or perhaps more effectively appended to the results relayed on the electronic medical record. Another example is the urinary antigen testing for legionella. Since sputum culture techniques are cumbersome and not widely available in community hospital laboratories, the urinary antigen is frequently relied upon to identify this organism. Frequently the clinicians are not aware that this test identifies only *Legionella pneumophila* 1, so that if another species of legionella is present this could represent a "false negative". Armed with this information a clinician may proceed with additional testing orders, and broader presumptive treatment, if the patient presents with a compatible clinical syndrome.

In Europe, Read et al. noted that clinical microbiology is incorporated in the training of infectious disease physicians but the duration of their training varies (Reed et al. 2011). Infectious diseases fellowship directors in the United States have long seen the value in having their fellows spend some time side by side with the microbiologist evaluating the daily cultures. The extent of this U.S. training is variable as well and not specifically proscribed by the accrediting body, the Accreditation Council for Graduate Medical Education. Regardless fellows should learn the pitfalls of inappropriate culturing as well as the techniques and methods used for optimal diagnosis of a particular disease or pathogen.

I. Unusual Clinical Scenarios

If a physician is faced with an unusual clinical scenario, it's important to discuss this with the microbiologist before samples are submitted. This will allow the microbiologist to prepare for the optimal cultures and other methods before the specimens come to the laboratory. There may be occasions where certain testing methods are not available in the hospital laboratory, the microbiologist can help advise and prepare transport media so that specimens can be sent to a referral state laboratory or to the Centers for Disease Control. Some examples of this are blood samples for Ebola for a patient coming from West Africa, special culture methods for Neigleria fowlerii in a patient with meningitis that occurred after waterskiing in a lake, or culturing for Mycobacterium chelonei from skin lesions of a female who developed multiple leg lesions after a pedicure.

Since physician specialists may have different needs depending upon the types of organisms they may potentially encounter, knowledge of available and pertinent tests is imperative for the appropriate and timely treatment of their patients. A discussion with the microbiologist is essential to avoid pitfalls.

J. Unusual Numbers of Patients with Similar Exposures

If the clinician sees a cluster of patients presenting with a similar clinical feature, and a common exposure it is essential to notify the microbiologist of this occurrence. The microbiologist can then recommend specific testing that would help in the diagnosis. If these cases were to result in an outbreak investigation, then it is important to order all the appropriate tests from the outset, particularly on the index case and potential sources. Additionally, such a cluster could potentially represent a bioterrorism event and therefore it would be imperative for the clinician to notify the laboratory so they can be extra vigilant in the protection of their staff from organisms that represent biohazards (such as tularemia, anthrax and the Ebola virus). The microbiologist can also help the clinician decide the appropriate order, the appropriate collection methods, and the appropriate transport media for successful outcomes.

It is vital to include the microbiologist in any outbreak investigation team. Collaboration with the clinician will allow more appropriate testing of symptomatic and asymptomatic patients with common exposures. On occasion, the laboratories are called upon to do environmental cultures requiring surface (e.g., *C. difficile*, MRSA), air sampling (e.g., aspergillus) or water sampling (e.g., pseudomonas, legionella). The microbiologist can help direct methodology for acquisition of these specimens. If the hospital laboratory is not licensed to do environmental sampling, they may direct the clinician to other laboratories that can effectively perform the necessary testing.

K. Process Improvement

If a process improvement involves management of patients with a possible or proven infectious disease, it is necessary to involve the microbiologist. This will allow the microbiologist to prepare for a new stream of testing or to suggest alternative methods for screening and testing. One example of this took place at the author's community hospital and involved the adoption of the preoperative screening of patients before total joint replacement. The team was interested in screening both for methicillin sensitive (MSSA) and methicillin resistant (MRSA) strains of *S. aureus* since several infections in the preceding year were with the sensitive strain of staphylococcus. The microbiologist recommended purchasing a special split plate that allowed nasal screening of both simultaneously. This avoided the ordering of two separate tests on two separate culture plates, which was more cost effective. Many ordering physicians were also not aware that the screening for MRSA did not effectively pick up MSSA strains, so they could have interpreted a negative result as fully negative had we not had this discussion in this committee. Infectious disease physicians, while a valuable member of the team in identifying disease process and in the prevention and treatment of infectious diseases, may not be aware of special laboratory methods, techniques and new microbiology developments so collaboration is essential. This communication allows all of the members of the team to be more informed and leads to more successful outcomes.

At J.T. Mather Hospital a 2008 campaign was initiated at their 248 bed community hospital over 7 years to reduce MRSA. They used PCR technology to initiate a rapid active surveillance of high-risk groups. While the cost of the testing and the increase in personnel was $650,000, the savings due to fewer infections was estimated to be over $2 million. The director of this program, Dr. Uettwiller-Geiger, was quoted as saying, "all stakeholders saw open communication and collaboration as being paramount to achieving success" (Uettwiller-Geiger 2015).

Summary

Excellent care for patients with infectious diseases depends upon collaboration and good communication between the clinician and the microbiologist. Gaps in this could lead to inaccurate diagnoses and prescription for incorrect or needlessly broad antibiotic regimens. Ultimately poor clinical outcomes and increased antibiotic related complications could result. From a system standpoint, communication between the microbiologist and clinicians at all levels of training will lead to improved infection control, better antibiotic stewardship and more cost effective care. Currently available rapid diagnostic methods will become even more available and less costly to resource-limited hospitals. The guidance of the microbiologist on the infection control team, the antibiotic stewardship committee and other process improvement committees that strive to reduce infections is invaluable, as they will lead to more evidence-based and thoughtful protocols in these areas. Additionally, communication between the clinician and the microbiologist will make the output from the laboratory more reliable, user friendly, useful, and satisfying while providing a safer environment in which to work.

References

Alahmadi, Y.M., M.A. Aldeyab, J.C. McElnay, M.G. Scott, F.D. Elhajji, F.A. Magee et al. 2011. Clinical and economic impact of contaminated blood cultures within the hospital setting. J. Hosp. Inf. 77(3): 233–236.

Al Saleem, N. and K. Al-Surimi. 2016. Reducing the occurrence of errors in a laboratory's specimen receiving and processing department. BMJ Quality Improvement Reports 5(1): 1–4.

Barenfanger, J., J. Benté and G. Havener. 2009. Optimal performance for clinical microbiologists and their interaction with infection control staff. Clin. Microbiol. Newsletter 31(2): 9–15.

Baron, E.J., J.M. Miller, M.P. Weinstein, S.S. Richter, P.H. Gilligan, R.B. Thompson Jr. et al. 2013. A guide to utilization of the microbiology laboratory for diagnosis of infectious diseases: 2013 recommendations by the Infectious Diseases Society of America (IDSA) and the American Society for Microbiology (ASM). C.I.D. 57: 485–488.

Beavers, T. and J.G. Wheeler. 2010. Collaborative medicine: weaving the microbiology laboratory into clinical practice. MLO Med. Lab. Obs. 42(10): 20–2.

Georgiou, A., M. Prgomet, G. Toouli, J. Callen and J. Westbrook. 2011. What do physicians tell laboratories when requesting tests? A multi-method examination of information supplied to the microbiology laboratory before and after the 2015 introduction of electronic ordering. Int. J. Med. Informatics 80(9): 646–654.

Ho, E.C., G. Chawdhary, A. Khan, S.G. Jones and M. Simmons. 2010. Ear microbiology reports: a need for better communication with the microbiologists. Eur. Arch. Oto-Rhino-Laryngology 267.4: 501–505.

Huang, R., D.J. Guervil, R.L. Hunter and A. Wanger. 2015. Lower antibiotic costs attributable to clinical microbiology rounds. Diag. Micobiol. and Inf. Dis. 83: 68–73.

Magill, S.S., J.R. Edwards, W. Bamberg, Z.G. Beldavs, G. Dumyati, M.A. Kainer et al. 2014. Multistate point-prevalence survey of health care-associated infections. NEJM 370(13): 1198–1208.

Mannis, M.J. and E.J. Holland. 2016. Practical ophthalmic microbiology for the detection of corneal pathogens. pp. 123–131. *In*: Cornea, St. Louis, Missouri, 4th edition. Elsevier Press, Inc.

Moore, J.S. and R.J. Koerner. 2015. In the era of the 24 h laboratory, does communicating gram stain results from blood cultures flagging positive outside of conventional working hours alter patient management? J. Clin. Path. 68.11: 938–941.

Murray, P.R. and F.G. Witebsky. 2009. The clinician and the microbiology laboratory. pp. 233–265. *In*: Mandell, G.L., J.E. Bennett and R. Dolin (eds.). Principles and Practice of Infectious Diseases. Churchill Livingstone, Philadelphia, Penn., USA.

Nishi, I. and Y. Hidaka. 2016. The role of the microbiology laboratory in healthcare-associated infection control. Rinsho byori. Japanese J. Clin. Path. 64(3): 338.

Peterson, L.R., J.D. Hamilton, E.J. Baron, L.S. Thompkins, J.M. Miller, C.M. Wilfert et al. 2001. Role of clinical microbiology laboratories in the management and control of infectious diseases and the delivery of health care. C.I.D. 32: 605–610.

Read, R.C., G. Cornaglia and G. Kahlmeter. 2011. Professional challenges and opportunities in clinical microbiology and infectious diseases in Europe. Lancet Inf. Dis. 11: 408–415.

Sautter, R.L. and R.B. Thomson. 2015. Point-counterpoint: consolidated clinical microbiology laboratories. J. Clin. Microbiol. 53(5): 1467–1472.

Schofield, C.B. 2006. Preventing errors in the microbiology lab. MLO: Medical Laboratory Observer 38(12): 10–12.

Self, W.H., T. Speroff, C.G. Grijalva, C.D. McNaughton, J. Ashburn, D. Liu et al. 2013. Reducing blood culture contamination in the emergency department: an interrupted time series quality improvement study. Acad. Emerg. Med. 20(1): 89–97.

Spencer, M., D. Uettwiller-Geiger, J. Sanguinet, H.B. Johnson and D. Graham. 2016. Infection preventionists and laboratorians: Case studies on successful collaboration. AJIC 44: 964–968.

Uettwiller-Geiger, D. 2015. How technology contributed dramatically to decreased HAIs and delivering high value outcome. Presented at: Clinical Laboratory Management Association: Increasing Clinical Effectiveness. Clinical Laboratory Management Association General Session: Orlando, Florida.

Var, S.K., R. Hadi and N.M. Khardori. 2015. Evaluation of regional antibiograms to monitor antimicrobial resistance in hampton roads, Virginia. Ann. Clin. Microbiol. and Antimicrobials 14(1): 22.

Zimlichman, E., D. Henderson, O. Tamir, C. Franz, P. Song, C.K. Yamin et al. 2013. Health care-associated infections: A meta-analysis of costs and financial impact on the US health care system. JAMA Intern. Med. 173(22): 2039–2046.

Index

N

O

P

Q

R

S

T

U

V